Praise for *The Nature of Drugs*

"Legendary chemist, nuanced psychonaut of molecular structure-activity relations, deep thinker on issues of societal policy, engaging storyteller, inspirational teacher, and all-around good human being—Sasha Shulgin takes us on an alchemical educational journey as if we were sitting there as students in the class from which this text arose. What a gift!"

— David E. Presti, Professor of Neurobiology, University of California, Berkeley
   Author of *Foundational Concepts in Neuroscience: A Brain-Mind Odyssey*
   and *Mind Beyond Brain*

"If you're curious about any drug, from caffeine to LSD, this is the book for you. What an absolute treat to learn from the best, to have Professor Shulgin as your personal instructor, with all of his charming, self-effacing asides and his witty encyclopedic knowledge on display."

— Julie Holland, MD
   Author of *Good Chemistry: The Science of Connection from Soul to Psychedelics*
   Editor of *Ecstasy the Complete Guide* and *The Pot Book*

"These course lessons are pure Sasha: enthusiastic, surprising, tangential, goofy, and shockingly knowledgeable. But they also give us something that remains terribly rare, even at this late date: a kaleidoscopic approach to the problems and possibilities of drugs that is at once pragmatic, visionary, and genuinely inter-disciplinary. Once again, Shulgin proves himself a magnificent spirit as well as a magnificent mind. I learned a lot, and enjoyed myself tremendously."

— Erik Davis
   Author of *High Weirdness: Drugs, Esoterica, and Visionary Experience in the 70s*

"The late, great Sasha Shulgin, a.k.a. "Dr. Ecstasy," was a pioneering psychonaut who designed a dizzying array of psychoactive substances in his mad scientist laboratory hidden in the hills east of San Francisco. These meandering musings are the surviving record of a course he taught in the 1980s, the decade of "Just Say No." They reveal the light-hearted human side of a chemist who—when it comes to drugs—advises us to 'Just Say Know.'"

— Don Lattin
   Author of the psychedelic trilogy: *The Harvard Psychedelic Club, Distilled Spirits,* and *Changing Our Minds*

"This book originated from a transcription of lectures given by Sasha Shulgin at San Francisco State University in 1987. In that respect, some of the content is dated. But for those who never had the opportunity to meet Sasha or hear him lecture, the transcriptions reflect Sasha's vibrant lecturing style. Indeed, if you did know Sasha, you can almost hear him speaking in the words of this book. It is replete with the kinds of anecdotes and analogies that were characteristic of Sasha's speaking style. He talks about what drugs are and their sources, different routes through which they enter the body, what they do in the body, how they leave the body, in addition to presenting information about various drug classes. Sasha sprinkles his lectures with questions about ethical and legal issues around drugs, and in general asks the reader to think deeply about some of the moral issues confronting us about drugs today."

— David E. Nichols, PhD
    Distinguished Professor Emeritus, Purdue University

"These transcriptions of lectures by Alexander Shulgin sparkle with the brilliance and wit of a pioneering researcher of the chemistry and effects of psychoactive drugs. Those fortunate to have known Sasha Shulgin will recognize his voice in these pages and take pleasure in listening to him share his wealth of knowledge and personal experience with magical molecules. A great read!"

— Andy Weil
    Author of *From Chocolate to Morphine* and *The Natural Mind*
    Director of Andrew Weil Center for Integrative Medicine at the University of Arizona

"Sasha was a beloved friend, a brilliant chemist, and an inspiration to everyone he met. This series of lectures captures the energy, warmth and irreverence that so typified his character. His underlying philosophy and approach to research remained forever youthful—a passion for knowledge; the need to constantly question established dogma, authority, and one's own previous assumptions; the vital nature of individual freedom of choice; and the pursuit of knowledge for the sheer, child-like thrill of it. Though Sasha is with us no more, we are lucky to have the ever-radiant Ann, the Shulgin Farm, and this beautiful book, *The Nature of Drugs*, embodying his memory."

— Amanda Feilding
    Executive Director of the Beckley Foundation

"Dr. Alexander 'Sasha' Shulgin was a pioneer, giving us a pharmacopoeia of hundreds of compounds—many of which have not been thoroughly evaluated to this day. Yet, Sasha was far more than a chemist; Sasha was a philosopher, a mystic, and a gifted teacher. He was able to present the 'boring' subject matter of medicinal chemistry in a way that was compelling and fascinating. And he didn't just teach chemistry; he placed the chemistry that he was teaching into the context of society, law, and history. Although Sasha taught widely and in many venues, nowhere is his teaching better represented than in the course he taught at San Francisco State University in 1987. Anyone with interests in science, chemistry, psychedelics, history, or philosophy, upon reading *The Nature of Drugs* will be rewarded with an incredibly fascinating an enriching experience."

— Dennis McKenna
   Editor of *Ethnopharmacologic Search for Psychoactive Drugs: 50 Years of Research*
   Founder of The McKenna Academy of Natural Philosophy—
   A 21st Century Mystery School

"Alexander Shulgin was many things, but first and foremost he was a teacher: he taught students, law enforcement, physicians, and eventually the world through the publication of his books PiHKAL and TiHKAL. This is Alexander Shulgin at his sharpest and most passionate. Emboldened by the emergency scheduling of MDMA and the passage of the Federal Analogue Act only three months previously, he offers a series of discursive lectures on medicine, pharmacology, human physiology, philosophy of science, astrology, alchemy, law, and linguistics. This text is a precious opportunity to attend a class taught by one of the great scientific thinkers of the 20th century and an indispensable primer for understanding the immensely complicated subject we call 'drugs.'"

— Hamilton Morris
   Documentarian and Chemist
   Creator and Director of *Hamilton's Pharmacopeia*

— The Nature of Drugs —

The Nature of Drugs

History, Pharmacology, and Social Impact

VOLUME ONE

ALEXANDER SHULGIN

FOREWORD BY Mariavittoria Mangini

TRANSFORM
PRESS

Co-published by:
Transform Press | P.O. Box 1152, Berkeley, CA 94712
Synergetic Press |1 Bluebird Court, Santa Fe, NM 87508
& 24 Old Gloucester St. London, WC1N 3AL England

Library of Congress Cataloging-in-Publication Data is available.

ISBN 9780999547212 (hardcover)
ISBN 9780999547229 (ebook)

Cover & Interior Illustrations by Donna Torres
Cover design by Ann Lowe, Amanda Müller, Brad Greene
Book design by Brad Greene
Managing Editor: Amanda Müller
Transcribed by Melitta Konrádi
Transcription Corrections: Keeper Trout
Printed by McNaughton & Gunn

TABLE OF CONTENTS

Acknowledgments . . . . *xiii*

Publisher's Note ~ by Wendy Tucker . . . . *xv*

Foreword ~ by Mariavittoria Mangini . . . . *xvii*

Introduction ~ by Keeper Trout . . . . *xxv*

LECTURE 1: Course Introduction 1

Overview and Introductory Discussion

    Sasha's Introduction

    "What is a Drug?"

Definitions from Lectures 1 and 2

LECTURE 2:

Lecture 2 is not presented here as such. The tape for the first half of Lecture 2 is missing. Accordingly, Sasha's written notes were used to interweave the discussions of definitions in Lecture 2 into the latter part of Lecture 1 and into Lecture 4 as he had done.

LECTURE 3: The Origin of Drugs 63

Overview

Opening discussion: Sasha's view on drugs, laws, power, and social control

The Origin of Drugs

    Human Development and Origins

        Alchemy

        Chemotherapy

The Origins of Drug Scheduling, Enforcement and How Drugs are Scheduled

Definitions

LECTURE 4: **The Plumbing of the Human Body**.... 109
Overview
Pharmacokinetics and Pharmacodynamics
    How Drugs Get into the Body
    Drug Administration Methods
Biotransformation:
    The Half-Life of Drugs in the Body
    Metabolic Interactions of Drugs

LECTURE 5: **More Body Plumbing &**
the Nervous System 141
Overview
Biotransformation Continued:
    Absorption, Excretion, and Routes
    Administration Methods and their Effects on Drugs
The Nervous System
    The Central Nervous System
    The Autonomic Nervous System
    Nerves, Structure, and Function

LECTURE 6: **Drug Action** 195
Overview
    Drug Treatment Categories
    Sasha's Drug Treatment Classifications
    Review of Drugs and The Nervous System

LECTURE 7: **Memory & State of Consciousness** 233

Overview

States of Consciousness

Memory

Changes in Consciousness without Drug Intervention

Mental Illness

LECTURE 8: **Research Methods** 265

Criminalistics and Drug Testing

Defining the Research Question

Devising a Hypothesis

Clinical Drug Trials

Human versus Animal Testing

Sasha's Lecture 3 Notes, 1987 301

Due to an abundance of supplementary material from Lecture 3,
  Sasha's original typewritten notes will follow Lecture 8.

Reflections from a Former Student ~ by William Leonard Pickard 319

Afterword ~ by Paul F. Daley 329

References 333

Index 335

ACKNOWLEDGEMENTS

The Nature of Drugs has been an effort involving many people spanning some years. The overarching vision, input, and guidance came from Ann Shulgin and her daughter Wendy Tucker. Scott Bodarky digitized the tapes and worked to clean up the audio quality as well as the mountain of ephemera and media in the barn. Transcription was begun by Stacy Simone while she was working digitizing the wealth of photos. The transcription was taken over and completed through lecture 8 by Melitta Konrádi. Sylvia Thyssen helped to edit and create a readable text. It was proofed and corrected by Keeper Trout who additionally filled in missing elements by re-listening to the audio files and using Sasha's written notes. The flow is therefore not exactly the same as the transcribed content, but the words are Sasha's. The text was further proofed by Jon Singer, Wendy Tucker, and Robin Donovan.

This book is a time capsule.

When Sasha taught this class, computers were in use but most of his research was done through periodicals, journals, and books. He went to the library. You'll hear him referencing his favorite journals, and he will describe how to find the information desired, but of course this has changed a great deal since then. This is true too with the information he gives about patents and laws; it is not up to date.

We ask that you keep this in mind as you read this book, and to understand that regardless of these details, his advice and wisdom about *how to do research well* is still true. If Sasha were alive teaching this class today, he would probably tell people that they must be even more diligent than ever to weed through all of the misinformation there is on the internet. Research may be easier to do in some ways, but finding what is real and true takes more work, more skill, more objective and critical thinking, and is probably more difficult than before.

Enjoy this first volume of *The Nature of Drugs*!

— Wendy Tucker
Publisher, *Transform Press*
February 2021

There are many aspects of Sasha Shulgin's life and work that will likely be familiar to readers of this book. His best-known works, *PiHKAL* and *TiHKAL*, written with Ann Shulgin, combine the stories of their courtship and marriage with chemical synthesis information from private notes on psychedelic compounds. Working in his lab, in a shed in the yard of the Shulgin Farm, his family home, Sasha independently created more than 200 psychoactive substances. Although his research papers describing the hundreds of unique chemical compounds that he synthesized have been widely published, and his work is of unquestioned importance, his research was not subsidized by a university, a government research facility or an industrial sponsor. Instead, he supported his work and maintained his independence from potentially censorious influences by consulting, lecturing and teaching. The Farm today is the repository of a treasure trove of research documents and reports, lectures, journals, letters, and photographs that comprise Sasha's scientific library and personal papers, which are now being explored and digitized. This book, which represents the transcripts of the first semester of a pharmacology class taught by Sasha in 1987, is a part of that effort.

In approaching the material presented in *The Nature of Drugs*, it is important to note that the social environment in which a contemporary reader encounters these lectures is significantly different from the one that prevailed when they were presented. In 1987, this country was in the midst of a moral panic about drug use. The use of plants and chemical compounds for the purpose of consciousness alteration was generally considered to be a criminal act. Although research indicated that many users of illegal psychoactive drugs were able to function effectively and undetectably in society, most public policy of that time presupposed that such use would inevitably have demonstrable negative consequences.

Independent Research (handwritten margin note)

At the time that these lectures were presented, the popular and the legal formulations of the use of illegal drugs allowed, for the most part, for only two patterns: abstinence and abuse. Any illicit drug use was defined as abusive, and moderate use was believed to be an unstable pattern, which might at any time deteriorate into uncontrolled use or drug addiction. Studies of drug users often failed to differentiate between different patterns of use, employed imprecise and inaccurate terminology in describing levels of use, or made no attempt to describe patterns of moderate use. The majority of theories that explained drug use and described the drug user did so in such negative and pathological terms that it seemed mysterious that any drug users survived at all. Sasha was one of the minority of drug experts who recognized the existence of a large number of experimental or occasional users who did not present any serious problem in terms of morbidity and mortality. On the contrary, he understood that drug users might value their experiences for many different reasons that did not arise from pathology, and that most drug users do not become abusers or addicts.

Ironically, the existence of patterns of moderate use was most publicly recognized by William Bennett, "drug czar" of the first Bush administration, who acknowledged the possibility that experimental, infrequent, or even regular non-compulsive use of illicit drugs might have few detectable effects on the health, work, families or social lives of some users. Nevertheless, in Bennet's 1989 National Drug Control Strategy, he singled out these "non-addicted casual users" for his strongest opprobrium, calling them "potential agents of infection for non-users," presumably because they did not fit his description of drug users as "inattentive parents, bad neighbors, poor students, and unreliable employees" whom no one would wish to imitate.

With the ascendance of political conservatism during the Reagan administration, problems that might be associated with the use of prohibited drugs had come increasingly to be viewed as resulting from moral, spiritual or biological defects of individuals, rather than as the product larger of social or environmental problems. The remedy for these dysfunctions was greater social control, as opposed to social welfare. Cultural conservatives determined that they could use drug prohibition as a legitimate source of control over unruly elements: minorities, youth, "aliens" and cultural liberals; the "dangerous classes" that seemed to be getting out of control in the 1960's and 70's.

Drug prohibition became a powerful exponent of the projects of the cultural right. It displaced concern for social conditions such as poverty, lack of educational opportunity, racism, unemployment, and a deteriorating social safety net; and concentrated explanation of them on the deficiencies and weaknesses of the affected individuals. For groups that were experienced by conservative elements in society as disorderly, rebellious, and disrespectful of authority, opposition to their non-conforming behaviors was seen as a reaffirmation of social hierarchies and traditional moral values. Control of the use of drugs that might be favored by these groups was used to provide the justification for increased social control and drug prohibition was linked with expressions of racial and ethnic intolerance.

Sasha challenged this prohibitionist stance both for its provincialism that specified a limited range of acceptable interests and experiences and for its paternalism that surrendered autonomous decision making in return for a promise of security and safety. He drew critical attention to the substitution of lies, distortions and fallacies for history and scientific evidence that was characteristic of legalistic views of drugs and drug use. By contrast, he offered potential drug users a singular position: learn the facts, then make an informed decision for yourself. Rather than trying to control citizens' choices, or resorting to hyperbolic fear-mongering, he advocated a pragmatic alternative: scientific and realistically grounded education. This was drug education that was not directed solely at prevention of use, but which also provided those who chose to use with information that encouraged moderation, appreciated the legal consequences and social realities of drug use, and was based upon science.

Much of this first series of lectures is devoted to Sasha's view of teaching and learning. He encouraged students to listen to the emerging "music" that the interplay of his planned lecture outlines, inquiries and offerings from class members, and his own stream of thoughts produced; and he discouraged excessive notetaking as a distractor from the experience of in-person interaction with the class material. He wanted students to get the feel of his work as a chemist, but also as a philosopher and an artist. Drugs, he claimed, were incidental to his presentation of ideas about free choice and informed citizenship. They provided an occasion for Sasha to talk about what he claimed really went on in this class: an experience of learning that was designed to equip us to have freedom of choice and to retain our personal power of discernment in our decisions and actions. Sasha was an early endorser of "just say know" as an alternative to "just say no."

Those of us who were fortunate to spend time in conversation with Sasha can almost hear his voice as we read these transcripts. He was an engaging teacher, masterful and authoritative about his subject, but ready to acknowledge ambiguities and areas that were outside his expertise. He was quick to grant that there were exceptions to his expert knowledge and that ten percent of his ideas would be likely disproven. His expressed wish in these classes was for the students to listen to the "music of chemistry," which he saw as a creative exercise.

The incontestable principles of chemical structure served as a launching point for discussions about societal issues and controversies. In this series of lectures, which was intended for a general audience with no particular background in chemistry, Sasha prepared his listeners for the detailed discussions of specific drugs that would be presented in later classes. He covered basic anatomy and neuroanatomy, physiology and neurophysiology—our "plumbing and wiring" as he called them—pharmacological concepts such as pharmacokinetics and pharmacodynamics, definitions used to describe drugs and their effects, and a chronology of drug law and policy. By introducing the pharmacokinetic processes that explain the way that the body reacts to drugs and the pharmacodynamics of receptor effects and chemical interactions, Sasha opened a greater access to wide-ranging discussions of drug action for students who might lack a background in these areas. More importantly, these lectures provided an opportunity for Sasha to present his opinions, convictions and principles in a context that attracted students with an interest in the role of psychoactive substances in human experience.

Sasha's teaching sometimes followed an outline or notes and sometimes, by his own admission, "just rambled around." In this lecture series, he laid the groundwork for further explorations that would come later, equipping his audience to participate with him in some wide-ranging conversations about individual drugs, what they do, and how we think and legislate about them. In order to do this, Sasha presented foundational knowledge that would permit his students to formulate pertinent questions and to follow him in his rambles. This background of understanding was meant to allow class time to be an interactive learning environment, where both students and instructor could engage creatively with the subject matter. Sasha used a broad definition of "drugs" which included "those things that influence a living organism or behavior," and he provided a picture of where drugs come from and where they go, their actions, their risks, and their virtues. This became the basis for his later lectures not only on specific drugs, but also on topics that relate to a "wilder territory" that embraces as drugs such things as smog, radioactivity and pesticides.

The importance of Sasha's digressions and asides can't be overestimated. He provides practical examples that illustrate important concepts in a way that is in keeping with his position about note taking—that it distracts from listening appreciatively to the material being presented. He uses his position as an experienced researcher to discourage attempts to "prove" a hypothesis, and to remind his students that science advances by discovering that previous hypotheses are incorrect. He emphasizes the importance of asking the appropriate research question, and of designing inquiry as a quest to disprove hypotheses, since an experiment that proves a hypothesis is impossible to devise. He also delves into areas of philosophy and policy, and gives invaluable advice about the nature and conduct of research. Sasha was acutely aware of the way that powerful persons, governments, and agencies may be more committed to avoiding any admission of error than to moving from ignorance to enlightenment and that there must be places where the laws can be broken in order for society to test their merit.

Sasha's comparison of drug education to sex education gives a hint of the perspective that led him to describe his chemical diagrams as "dirty pictures." The idea of both is to equip the listener to make informed decisions when the opportunity and the inclination to engage in certain behaviors may coincide. He was not reticent about including his views about social ethics in his lectures, but he did not advocate for or against drug use. He refuted the conventional ideas about abuse of drugs that define "abuse" as use of any forbidden drug, lack of permission from a specific professional prescriber for use, acquisition of drugs through unsanctioned channels, or use of drugs in ways that may introduce an element of danger. Sasha held to a definition of drug abuse that has nothing to do with illegality or medical supervision: as long as drugs can be used in ways that don't interfere with social functioning or mental and physical health, their use is not here considered to be abuse. In this more complex view of what constitutes abuse, the relationship of an individual to a substance is the crucial concern.

In order to have a foundation for following Sasha's lectures, and to keep up with his observations, stories and replies to their individual inquiries, students were encouraged to use the class' textbook to gain a basic vocabulary and understanding of the various kinds and classes of psychoactive drugs. The textbook *From Chocolate to Morphine,* by Andrew Weil and Winifred Rosen, was first published in 1983, a few years before these lectures were presented. It is still in print almost 40 years later, and the ideas that made it an obvious choice for Sasha to select have come to enjoy greater acceptance: that the desire to alter consciousness is an innate, normal human drive, and that problems with drugs arise from poor relationships with drugs, not from the inherent characteristics of the drugs themselves.

At the present moment, Sasha's devotion to the presentation of truths backed up by scientific and historical evidence presents a refreshing contrast to the encroachments of the current war on science. He saw himself as a truth seeker, rather than an advocate for or against drug use, and his work is an example of a cognitive lust: an intense desire to learn and know everything that there is to know about a fascinating subject. In his case, the inspirations for this enthusiasm were drugs that could cause not only visual and sensory changes, but could also modify and influence brain function. His interest was in drugs that "turn the mind;" psychotropic substances that can cause changes in perception, attitude, or point of view, and sometimes expand one's mental and emotional horizons or provide access to one's interior universe.

Sasha could be persuaded to admit that he was an alchemist. He approached chemistry as a sacred art and was mindful of the way in which the practical and the philosophical, the esoteric and exoteric, intersected in his work, as they did in classical alchemy. Historically, only a few practitioners of the arts he practiced rediscover this ancient truth. He viewed alchemy as a form of meditation, and chemistry as an art that was exactly like the composition of music, or the creation a painting: the putting of things together that had never been together before.

Sasha's practice was informed by the realization that the essential alchemical work was to understand oneself. While he delighted in the practical work of the laboratory, he recognized that the transmutation that is sought in the alchemical quest is a spiritual regeneration of the practitioner, in an evolution from ignorance to enlightenment. He was grounded in the real and tangible but respectful and curious about the mystical and intangible. Sasha mused about the possibility of embedding one's character in the substances that one works with, in a way that could be recognized by others. Beyond earth, water, fire and air, the elements of which everything is made, he was curious, as were the classical alchemists, about the quintessence, the fifth element, the spirit in matter, the soul that puts all the rest into place. In his work, Sasha Shulgin was able to achieve one of the goals of classical alchemy: to manifest spiritual forces for transformation in a material form. This class was a vehicle that he used to expose his audience to the ideas that shaped his work, and which lent a quintessential character to the substances that he created and synthesized. For us who have had the benefit the transformative power of Sasha's discoveries and innovations, these lectures give a unique glimpse of his remarkable mind at work.

— Mariavittoria Mangini, PhD, FNP
Cofounder, Women's Visionary Council
December 2020

What you are about to read is much more than a series of class notes, it is a taste of Sasha at his most entertaining. The Nature of Drugs was a very popular class taught at SFSU by Sasha (Alexander T.) Shulgin. This book was from one of those classes, recorded in 1987. These recordings have been transcribed to enable the experience of being in his class, to get a taste of Sasha's wildly free-form and fast paced lecturing and speaking style.

Despite rolling through technical subjects such as pharmacology in general, pharmacodynamics, pharmacokinetics, metabolism, excretion, toxicology, and forensics; this was an introductory level course aimed at students who had no background in chemistry, and required no other prerequisites. This makes what Sasha wants to say very accessible to anyone. Sasha loved teaching this class because of the opportunity it provided him to share, not only the core material, but a bit of himself in hopes of captivating and influencing those whom he believed would become future professionals in medicine, chemistry, pharmacology, and forensic sciences. His philosophical views on drugs, life, sex, personal rights and freedoms, societal concerns, and legal constraints were all freely shared, along with advice to reject out-of-control authority politicizing any area of study, and learning how to ask the right questions. In short, how to perform good science.

Volume One of this three-part lecture series discusses: How a drug gets into the body, how it moves around, what it does, what happens to it, and how it gets out. In doing so, Sasha attempts, as he puts it, to present "what can be bad about drugs, and what is sometimes very good about drugs. Warts and all."

— Keeper Trout
   Author and Editor, *Trout's Notes*
   November 2020

Course Introduction

SASHA: All right. The name of the course is "The Nature of Drugs." It was originally going to be "Drugs and Society," which would be kind of a neat thing because you can go and tie everything together in a nice way, but some other department had already stolen the name and refused to give it up. They had to find something new and "Nature of Drugs" had not been used, and so that's what this will be. But it doesn't matter very much which name it has or through whose auspices it is authorized, since all of that will have little influence on the content of this course. I intend to cover the area of drugs in the broadest definition of the term, and the attitudes of society towards them.

There are absolutely no requirements for the course. It's nice if you've had chemistry, but I'm going to largely try and resist my big temptation to put great big hexagons on the board with wiggly chains out from the amino groups and methoxy groups and marvelous things like nitrogens. Because this is the heart: I really honestly believe that the knowledge of chemistry is the knowledge of one of the few disciplines that not many people are going to take issue with. You talk about a drug—we're going to talk about thousands of drugs during the course of the year—and you will often encounter some controversy.

Let's talk about mescaline, for example. You'll get controversy as to what it really does, and how it acts, and where it acts in the body, and what these receptor site things are that it acts at, and how it's metabolized, and whether something is it or not, and is it found in this cactus or that cactus. But you're not going to take issue that it has this structure.

So, the idea of a chemical structure allows one thing in this very pied area of so-called science to hang together. You can say what the compound is, you can say what its structure is, how to make it, and what its properties are, physically, chemically. What it does in the body is into the realms of art. But what it is in a test tube and in a beaker is one of the few really incontestable arguments.

I love chemistry as a focal point from which to say, "Here is a structure. What it does, I don't know. But maybe it does this and that." It's a nice starting spot. I'm going to resist it.

How many people have taken other scientific courses, let's say botany? Whee! Okay. I'll introduce some botany. I love it when a few have and most have not because then I can justifiably get into it a little bit further than I normally would.

I was going to introduce myself, but let me introduce you to me first. How many people have taken psychology? Oh, wow! How many people are in psychology as serious business? Well, some of them. At least you've gotten into the area.

How many people have taken caffeine? Only two? Oh my, my. I'm in the voting too. I've taken caffeine. How many people have taken caffeine? Now that's more like it. How many people have never touched caffeine? Okay, that's enough.

By the way, I'm a nut on vocabulary. I love vocabulary and I love using it. Sometimes I get a little bit carried away because I talk about the "hemioptus dysiptria" and I realize that dysiptria is not a common word in everyone's vocabulary. So, I think if this is going to be the size of the class it's really going to be neat because then it's going to be small enough that if someone says, "Hey, hold on. What is dysiptria?" I'll go into it and we're off on a tangent.

There's a textbook for the course. It's—what is it? *Chocolate to Morphine* by Andrew Weil. It's a nice one because it honest-to-god presents things as they are. I'm going to have a theme for this whole course called "warts and all." Namely, what is known about drugs, what is to be found out about them, what do they smell like, what do they taste like, what are the goods, what are the bads. Why is it so bad to use drugs? Why is it occasionally so good to use drugs? It's going to be an issue of talking about drugs and their properties. I am not going to champion their use and I am not going to espouse the argument of "just say no." Nothing along that line has anything to do with drug education. I love the analogy of sex education, which is exactly the same thing if you look at it, just from a different point of view. Sex education: "You've got to teach people just to say no." Well, this is fine and dandy because if they choose to say "yes" then they're out on their own because they've never learned anything outside of the very rudimentary males and females aspect of that.

Let me discover where you are, a little bit. Let me give you three possibilities and have everyone vote on them. The possibilities are: learning at home, learning at school, and learning from your peers. How many learned about sex at home? One, two, three, four, five, six, seven. Okay, that's about a third. How many people learned about sex at school in some class? Seven. Neat. Eight. How many people learned it from their peers? Eight. Okay, even. Just about even. I trust I shouldn't have asked how many people know about it. I assume that would have been a universal.

Same argument goes on drugs. You learn an awful lot about drugs from school. But a lot of what you learn is, you know, "Stay away from it." A lot of people will learn about drugs from home, but not as many as I think should because there is a lot of ignorance at home about what drugs are and what drugs do: "Good God! I smell something strange. You haven't been—? We gotta go talk to the minister." This kind of thing is a nice approach to morality, toward ethics, toward what is probably good behavior, and, in many peoples' eyes is the only right behavior. But it has nothing to do with drug education.

It's nice to learn it in school, but what you're going to find are the stereotypes. I saw this very beautifully in medical school where in the second year there was a course called "Pharmacology" that lasted for three quarters. In that class you learned all about pharmacology, which embraces drugs.

One of those three semesters of pharmacology was on CNS drugs: "This will turn 'em down. This will turn 'em up. This is a stimulant. This is a depressant. This is good for treatment of this. And that's good for the treatment of athlete's foot." Then, within that particular semester they had a one-hour lecture that dealt specifically with two topics: one was the psychedelic drugs and the other was smog. And it was all tucked into this short one-hour lecture of which forty minutes was on the psychedelic drugs, and the comments were, "They all cause a toxic psychosis. They are all the same. And substantially the best treatment is Haldol or one of the tricyclic tranquilizers," which are strong or heavy tranquilizers. We'll talk about strong and heavy in a moment.

[Directed to student] Yeah!

STUDENT: When was this?

SASHA: Ah, about fifteen years ago. I don't know if it's changed. I've not been there recently. But this was the attitude that was taken, and you'll still find, "You wanna find out really what is the problem about the use of marijuana? Well, let's go down and talk to our family physician. And he has, by golly, been through those courses and he knows what he's read, that it causes enlarging of this and decreasing of that and maybe increasing of something else." And you ask, "Well, why do people use it?" "Well, there's no really good reason." Nonsense! There's a perfectly good reason. They get high! [Laughter.] And you say, "Well, that's not part of our social ethic. Here, have a cup of coffee and wait until I handle someone else."

We have drugs all through our society. I'm going to start a tally. I wrote notes, shows how much I'm going to use them. We have caffeine. How many people have used alcohol? I'm amongst them. Okay. It's almost embarrassing to ask the question, "Is there anyone who hasn't?" Because the one Mormon, possibly the one Quaker, in the crowd doesn't want to put their hand up. It is all around us. How many people have used tobacco? How many people have not? This is a legitimate question. I have. I can't raise my hand. About two-thirds 'yes,' about one-third 'no.' The fourth I like to put on this list, how many people have used betel nut? One. Any more? Two. I have not. That's one I have not.

STUDENT: What is it?

SASHA: What is it? That's exactly why you're in the course. We're going to find out what it is. These four drugs constitute the four most broadly used psychotropic drugs in the world. Probably, either continuous use or occasional use or association with use of, in the sense of having used yourself, each of these drugs, more than one billion people. In a world of about six billion you're talking about one out of six having used one of these drugs.

Betel nut. We'll probably get to it when we talk about intoxicants and such. It's kind of a nice little thing. It's a little nut about the size of an acorn. It comes out of a palm tree. It's an *Areca*, the genus of the tree. *Areca catechu* is the species name. It's raised from the Philippines westward: the Philippines, the eastern coast of Southern Asia, throughout Southern Asia, into China, across into India and throughout India. It is raised throughout that entire area of the world. It is used by virtually all adults in that world. It is put either dried, different cultures use different ways, but usually it is either dried and used dry and smashed, or it's cut fresh. It's usually taken from the slightly unripe fruit. And it is often, not always, but often, wrapped in a leaf, known as the betel leaf, that comes from a vine belonging to the *Piperaceae* family. That name applies to both the nut from the palm and the leaf from the pepper. It's wrapped in this and shoved into the mouth, up against the gum and the lip. And it's left there. And if you add a little lime to it, make it a little bit basic, it tends to drain out colors and you'll find people often get reddish brown lips and gums and teeth get stained. In fact, black teeth or very dark teeth are a measure, as a beard is in China, of age and wisdom. You've been around a long time, you've used your betel a long time, you've gotten much wisdom from your passage through this vale of tears. And it's not considered a disfigurement, just part of the territory, like wrinkles and emphysema is from smoking. It's all part of the territory and it's a sign of belonging.

You have this as a major, major material. An interesting sideline, I was going to get into this when I got into tobacco and betel, but I'll get into it

now, which shows I'm totally disorganized, but I enjoy doing what I am doing. By the way, thanks for the question. Anytime questions come up, ask them. That's the way I know where you want to go. With betel, you have an alkaloid that's known as arecoline. Maybe I should start writing some of these down. By the way, you'll notice a tremendous resistance to indicate that it happens to be a tetrahydropyridine with a carbomethoxy group on it and an N-methyl.

(This is the sort of thing where I'd love to draw a dirty picture. I call them dirty pictures, things like hexagons with things sticking out and functional groups.)

Arecoline is an alkaloid. How many people know the term "alkaloid?" I'm going to be asking this several times. Not too many. Okay.

An alkaloid is a compound that comes usually out of plants—to a purist it comes out of plants—that contains a basic nitrogen. Usually with some complexity, but not always. It is a base and a caustic material that comes from plants, and most of the active materials, not all, but perhaps nine-tenths of the active materials in human beings that come from plants are alkaloids. Nicotine in tobacco is an alkaloid. Arecoline in the betel nut is an alkaloid. Alcohol is a non-nitrogenous material. Caffeine is a relatively neutral compound that contains nitrogen and is often classified as an alkaloid. Three out of these four major world drug materials are substantially alkaloid containing.

This combination is put together [referring back to betel nut use], it's then put into the mouth, and it's left there. It's always in the same place. It's like a cat chewing on the same tooth. Pretty soon a callus develops there, the tissue becomes hard. The erosion is stopped because of the change in the tissue nature, and it doesn't tend to burn or blister anymore. And when the goodies are depleted, and the person feels the slight euphoria and the fun and the pleasure of it is dropping off, in goes another. And when you go to bed at night, in goes one for the overnight. And in the morning out it comes and in goes a fresh one for the morning. It's like a quid of tobacco; but, this is betel nut. And it has been used for millennia throughout India, Southeast Asia, into the Philippines, and all through the islands in the Western Pacific.

Now, a problem has come up. This is completely apart from the introduction. A problem has come up in India in the last twenty years. In our culture, there's nothing wrong with shoving a little bit of snuff or a little bit of chopped up tobacco up in there and letting it go. You'll find some people will go through their entire day and night with a tobacco quid.

To touch just a little bit of chemistry, there's a part of the arecoline that is very, very responsive to what are called mercapto groups. In the body, there is a whole inventory of mercapto groups known as—oh, gosh, you have glutathione, you have acetylcysteine. Those in biochemistry could give me a half a dozen more. These little groups are very, very reactive groups.

In tobacco, the principal alkaloid is nicotine. You cure tobacco by putting in nitrites, just like you cure bacon by putting in nitrites. And these nitrites tend to give it an aging, a texture, a smell, a taste that makes your particular tobacco competitive. But this aging takes off the methyl group and puts on a nitroso group on nicotine, so you get what's called in the trade, nitroso nornicotine. This is probably one of the principal agents that is responsible for cancer, and the cancer that comes from tobacco. One whole hour, as you've probably looked ahead, is going to be devoted to tobacco, so I don't want to get too much into this. But this nitroso compound is probably neutralized in the body by these SH [sulfhydryl], these mercaptans, these glutathione and cysteine things. So it takes a long while for the cancer to express itself.

In arecoline, you have something that sops up SH groups and therefore sops up the very thing that makes you protected against tobacco. What has happened in the last twenty years in India is they've begun mixing tobacco and betel together. So you get the euphoria of the betel nut and you get the slight stimulation and the light headedness of the tobacco in the same package, in the same quid. So what's happening in the quid, the component of the betel nut that has the SH group scavenger property robs the body of the defense against the nitroso nicotine that comes in the tobacco, and in the last ten years the most prominent, the most numerous instances of cancer in India has been cancer of the mouth. It exceeds cancer of the lung, exceeds cancer of the bladder, and other cancers that have been associated with tobacco. Cancer of the mouth, usually of the gum or the throat, or of the jaws, a third level. Adding these two different drugs together for their goodies happens to compensate for the body's own defenses against each of them and here you have a superb example of a social problem—by that I mean preventable cancer—a social problem that comes directly out of drug use that is not only allowed, it's actually encouraged. It's advertised and it's promoted as being a very, very excellent experience.

So, you say, "You should tell them, just say no!" Well, no, you should begin saying there is an interaction here that's got a problem. Be aware of it. This course will be directed towards drug education. Drug education is a search for facts concerning drugs. As I suggested before, the current move to teach people to "just say no" may be good advice for some, and pointless for others, and it has both ethical and moral justification. But it has nothing to do with drug education.

Most of you have already been exposed to drugs, and most of you will personally decide if you wish to become exposed again in the future. The goal of this course is to provide specific information concerning drugs, as to their actions, their risks, and their virtues. And that's really what my role is, I'm a seeker of truth. I'm trying to find out what's there. I am not an advocate for nor an advocate against drug use. I have my own personal philosophies that have no business in here. You'll find that I am quite sympathetic with a lot of drugs that people say are evil and bad. But in truth, I want you to have enough information that you can decide for yourself whether this is something that's your cup of tea, quite literally caffeine, or whether it is something you wish to stay out of.

This is more or less my introduction. I have several bad failings. I jotted some down here to remind myself. One, I tend to lecture a bit too fast. This time I've kept myself under control. You notice we've gone at a very leisurely pace. [Loud laughter.] I'll try to keep it there.

I had a marvelous student in my Forensic Toxicology class at Berkeley a few years ago. She wore everything on her face, her affect was absolutely evident in everything she did. And when I said something she understood, there was this great big smile. When I went a bit too fast and used a word she didn't know, she couldn't help it, she went into tears and would quietly cry, her tears would actually run, and her whole face would cloud up like a storm. I used her as a bellwether. It was marvelous! I'd go lecturing along, I'd kind of glance over there occasionally, I'd see these tears rolling down. Hah! Slow down. Go back, go over it again. It was like talking to a jury as an expert witness when you're working on one person who looks vaguely intelligent and you say, "I'm gonna make the issue to that person who smiles and nods." As you explain a difficult point, you watch their face. If it frowns and is shaking side to side in confusion, go over your point with more care. If it smiles and nods up and down, move along. After all, that person will be your spokesperson in the jury room, so be sure that what you are saying at this particular time is being understood. I don't know the people well enough here to know who cries and smiles, and so I will try to go at a leisurely pace, which is not my usual way, and I will tend to slip into old habits, which is going roaring along. So someone holler and "Whoops! Would you spell that, would you write that down? What was the meaning of what you were just talking about?" I'll write it down. We'll try to do it at that rate. This whole first lecture's going to be a matter of introduction, one way or the other.

Questions in general. This is a small enough group that people can wave their hand and say, "Question!"

[Directed to student] Yes!

STUDENT: What was it that the betel nut robbed? What was it called?

SASHA: Okay, specifics. This is an exception to my rules. [Writing on board]. This is arecoline, an SH group, a mercaptan group, a thiol group, called mercaptan. An SH group is a functional group on a lot of molecules in the body. It's called mercaptan because it captures mercury.

Its origin is that it is something that grabs a lot of things including heavy metals. And including very, very reactive species. Mercaptan, mercury capturing. Hence the mercaptan group is one of the body's defenses and is one of the body's manipulations for handling things that are very reactive.

Cancer is generally formed by things that are very reactive in a very general sense. Things don't just go in and form cancer because they happen to have cancer written all over them in glowing letters. They go in because they have free radicals and they have reactive things. They go into the body and they glom onto things in a very easy way, and often they glom onto something that's very necessary for the normal regulation of the body. That regulation of a cell, whatever has been hit, is no longer there, and the cancer comes from that. The mercaptan, which protects the body in many ways or reacts with the nitroso compounds of T (for tobacco) will be captured by arecoline and not be available for its normal prophylactic role.

That lecture's down the line. I'll hit all this again later in a different way, but I wanted to get started in that way.

Okay. You've introduced yourselves. Let me introduce myself a little bit. My name is Alexander Shulgin and I have always responded to the nickname Sasha. My background is strange. I took undergraduate work at Harvard and at Berkeley. I have a degree in chemistry. I have a doctorate degree in biochemistry. I have spent some post-doctorate, post-graduate work in both medicine and psychiatry. It's sort of a weird collection of disciplines. My true love is pharmacology and things that affect the central nervous system. I have published some 150 papers and patents, many of which are concerned with the effects, in humans, of new or known psychoactive drugs.

My strengths as a lecturer are pretty straightforward. I am completely in love with the process of learning and am especially taken with any question to which I should know the answer but don't. I see myself as a truth seeker. I feel quite at home with elementary chemistry and am personally at peace with the actions of a number of psychotropic drugs.

I have experimented with many on single occasions and yet, at least at the present time, I use no drug chronically.

Bad habits: I dislike books to a large measure because very often I find that books tend to be written by people who want to impress you with how much they know. It's like people who used to write books in the technical area and are now working in computers. They write things that are manuals for how to use a new program, and all you get out of it is, "Gosh, he must know an awful lot to write a manual of this degree of complexity, but I don't understand it." One of my pet peeves is the introduction to a general subject using vocabulary that's jargon and is not at hand. What I'm going to try to do, at least in the first hour, is try to get a lot of these words out and really tell you what I think, and how I feel, their meaning is.

Drug education. I'll talk about it, about just saying no, all this sort of thing. I want to talk about drugs themselves. Everyone has a handout. If not, okay. Lean on someone and pick one up here afterwards.

[Directed to student] Yes, question.

STUDENT: Can I have you write the name of the textbook and the author?

SASHA: Okay, it's—I'll make a try with my spelling of *Chocolate*, C-h-o-c-o-l-a-t-e [Writing on the board], *to Morphine*. It has a subtitle, the something or other of somethings. It's *Chocolate to Morphine* and the author is Andrew Weil. Plus a coauthor whose name I do not know.

The book is totally lay, it is totally without complexity, and his thesis is really much of my thesis. I'm going to get into what's meant by drug abuse. To give you a preview of what he says, and what I feel, drug abuse is the relationship between a person and a drug—a drug and a person, together—in which the person does not have a good relationship with the drug. I'll use myself as an example.

I'm very familiar with a lot of different psychotropic drugs. Out of some 200, 250 psychedelics I probably have used 150 of them. I am familiar and I have not emerged with conspicuous brain damage so I think I can lay to rest [laughter from the class] that correlative. I have other sorts of damage, but I won't talk about that. [More laughter.]

The thing is I, for example, smoked for fifteen years and I smoked good and heavily. I stopped smoking. It was quite a strain. Believe me, that is one of the more addictive drugs (I'll get into the word addictive and what that means and why I will very rarely use it), one of the more psychological dependence-developing drugs. I stopped it. I stopped it the only way you could ever stop any drug use that you are not totally at peace with and that is by saying, "I choose not to use the drug." You may go into hypnosis, you may go into therapy, you may go into group encounters, you may have to lick dirty ashtrays, whatever. There are all kinds of approaches, aversion therapy to who knows what, often using a drug to break you of the habit of using a drug, which I consider to be sophistry at its worst level. The idea, "You're on heroin? Here! Go join our methadone clinic. You won't have to use heroin anymore." So you keep shooting up with, or swallowing in this case, methadone. It's ridiculous!

The idea is if you want to get off a drug or you want to get out of the habit, you want to get out of where you are, evaluate where you are and make one simple statement, "I choose not to be here." And that's it. You have stopped smoking, you've stopped drinking, you have stopped drinking coffee, whatever it is. What you do in the withdrawal process is come up with some of the most beautiful rationalizations you've ever seen in your life: "Who's running the show," "I want to drink," "If I want to smoke, I'm going to smoke." That's fine, that's fine. But once you've gone through that rationalization, you have answered the previous question negatively, you have not chosen to stop.

The addictive potential, expressed here as a poor relationship with a drug, is in all of us and it needn't be restricted to drugs. I was listening to Hal Lindsey's fundamentalist Christian radio program a few weeks ago, and got caught up in the program that followed it. On this "revival" session, a young spokesperson for the Church had a transformed drug abuser at hand, and they were unendingly vocal as to the virtues of finding Christ. I can only paraphrase the testimonial:

"I used cocaine. I destroyed my life with cocaine. I lost my job and my self-image with cocaine. But once at a moment of intense commitment, I said, "Jesus, I accept you," and from then on, I had no desire, no urge. And my wife, having seen the transformation, joined me in Jesus, and we are the most one-ness pair you could ever see. I will go anywhere, and talk to anyone, as to the virtue of Jesus over cocaine."

A commitment to an addiction, and with sufficient reinforcements towards that commitment, constitutes a conversion that is real. To exchange a total commitment to drugs for one for Christ might be seen not as a change in style, but simply a change in dependencies.

In this regard, drug abuse is a person's use of a drug with which they have a poor relationship. This is Dr. Weil's thesis, and I completely agree with him. I have chosen to stop smoking because I have a lousy relationship with cigarettes. I smoked two or three packs a day for about fifteen years. I gave up the habit cold turkey. Was off for about three or four years. Got involved with a very, very neat little romantic situation in France. Her husband was in Germany, but we were in Paris, which was [loud laughter] another whole story in its own right. And on the last night there she's taking the night train back through Belgium over to Cologne, we had a little Calvados in a little cafe we knew in the Sixth Arrondissement, she says, "Have a Gauloise." These are little French, blue cigarettes. Oh god, they're strong. No filter. A filter wouldn't even withstand what comes down that cigarette. They do have a filter now. So "Have a Gauloise." "No thanks, I don't wanna get—" "Aw, go ahead it's the last night." Okay, I'll have a cigarette.

Two days later, I was smoking two packs a day again. Right back into it, unbelievably. I said, "This is ridiculous!" It went on for a year. Then I stopped again, said, "That's it. I'm not gonna smoke anymore." I've not smoked since. I don't dare have a cigarette because I have a lousy relationship with tobacco. I think you have to evaluate your relationship with the drug and determine whether it's okay for you or not. I know a lot of people who can smoke a cigarette after dinner and that's it. I admire them; but it's not my cup of tea. One cigarette after dinner and

I'll go down to the store at two in the morning for a carton. [Laughter.] It's just that I know myself. Know yourself and establish that relationship. If you have a good relationship with the drug, I don't think that's drug abuse. Anyway, I'm getting out of my text.

This first lecture today is the introduction, getting to meet one another. Those who get panicked by the way I go, they don't have to show up again. I want to go into the history and practice not only of drugs but of medicine because it gives you a very good perspective on how recently we have become so sophisticated.

Just a few years ago, even a hundred years ago, there was no such thing as pharmacology, no such thing as drugs. For example, the word pharmacology didn't exist before 1890. You go to a medical curriculum in the nineteenth century and what we know as pharmacology they called *materia medica* because it was the study of things that came from nature that treated illnesses. Ten years earlier there was no concept of treating an illness with anything from nature, except for this for malaria and that for worms and that for amoebic dysentery. And something, by the way, for heart regulation and we all know these famous little plants.

But, all plants and drugs did was increase the body's resilience, or increase its energy, not treat a disease. The concept of disease connected to microbes or bacteria and such didn't exist 200 years ago, didn't exist 150 years ago. Illness was something to do with religion or something to do with your behavior patterns or your relationships with others and no idea of the causal agents that we now assign to disease origins.

I think very possibly we're going to make another change in another few years when we realize a lot of the diseases, a lot of the illnesses, we have that we blame on bugs actually stem from something upstairs in the head. I remember one very neat lecture I heard one time that impressed me. I never forgot the image that was given, which was to take a person and remove the clothes so you could see the body and recognize the person from the face and the torso and the legs and the arms, the entire person. And then recognize, "Ah yeah, that's Joe Brooks. I know him well." Then, in some way, remove every cell from that body

that belonged to that person, every cell that had his chromosome identity in it, so that actually, there was no person there, every cell was
somehow magically removed. You'd still recognize the person on the
basis of the fungi, bacteria, spores, the alien guests that wander around
our bodies continuously. You'd recognize the complete person: face,
torso, legs and arms. We are so laden with weird crawly things [laugh-
ter] inside and out.

Then why don't we come down with the roaring heebie jeebies continuously? We don't. You'll find every one of us has some tuberculosis
bacilli in us somewhere. We've got viruses in us that are pathological.
We manage to go along, have a cup of coffee, go out to the movie and
sleep soundly. A lot of this control comes from up here in the head,
that you choose (not consciously, unconsciously, but I don't want to get
into Jung here) for some reason and the time has come to come down
with a cold. Well, it turns out the day before you had to do something
you were trying to find a way out of anyway and *phht*, down you come
with a cold.

Go into the critical wards at the hospital where people are tottering
between life and death, where you have people in ICU intensive care
treatment. People maybe are at a very, very critical point of some life
process. The fever's here and the blood pressure's there and the liver
function's out there somewhere. You can often go and see the person and
know that person's going to live. The next person on the next bed—same
wiggle lines on the oscilloscope, the same weird chemistry at the foot of
the bed. You look at that person, that person has turned in their chips,
they are going to die, they have chosen to die at some level. This first
person's chosen to live at some level. And, by golly, three days later this
person's dead and this person's still alive and in recovery. Same pathology. The same physical state. It's all in the state of mind. I think at some
level you believe, you must believe, that where you go and how you end
up and how you find yourself comes much from your own choosing.
This idea of freedom of choice is to me a very, very necessary ally. It's a
very, very strong personal freedom.

In fact, I want to get into that. One of the things I have down here is I am a conservative. You may think that's a little bit silly, but it's not. I am very much a conservative and I've written here, that I'm profoundly disturbed by the thought of a pathological criminal wandering the street having been let out because there's insufficient evidence to hold him in jail. And I'm very disturbed by an encounter I had with a closed-minded student—I had one just last year—who had gone through a private school and had taken the entire Garden of Eden totally literally and had accepted on faith. The textbook had been supplied by the fundamentalist school board to that group. This disturbed me very deeply.

I'm disturbed by being on a 747 where the pilot staggers up there having had a wild night at a big pot party the night before; not quite baseline this morning, but he's still going to take the 747 out—that bothers me very deeply. But I much, much prefer to get mugged in San Francisco down in the Tenderloin than to have a student or a school system I feel is being changed by what I consider to be a one-sided and not totally balanced educational background. And I'd much rather take the risk of having a dangerous flight with a pilot than to invoke such things as the abrogation of the due process of law or invoke censorship or the robbing of freedom of choice and the curtailment of civil rights as is personified by urine tests and all that sort of thing. So, this is where I become a conservative. I am a very firm believer in the rights that are vested in the Constitution and I will do what I can to maintain those rights. Do not give away your rights for the sake of a little bit of social comfort.

Discussions always seem to come up that lead to an evaluation of the relationships between virtue and vice, but neither word has the slightest meaning in the absence of liberty. The American philosopher Denis Donoghue stated it exactly: "To choose vice is better than having virtue chosen for you."

This choice must remain, it must absolutely remain, it must be demanded by each of us continuously as a personal liberty. These changes are apparently all being made for the sake of a safe and bland society. Preventative detention, book burning, and urine tests are all intolerable to me for any reason whatsoever. I am a conservative believer in the Constitution, and in the personal freedom of choice, and in the right to be individual.

In fact, I have a beautiful, beautiful quotation right out of Ray Bradbury. How many people are familiar with Bradbury? It's within this generation, I hope. Yes. Good. *Fahrenheit 451*, dealing with book burning.

> "Now you see why books are hated and feared. They show the pores in the face of life. The comfortable people want only wax moon faces, poreless, hairless, expressionless. Worse still, the censors want the world's face to be their own. [This will apply very much in our discussion of what is good and what is bad in the way of drugs.] Classics and trash, good sense, nonsense, all reflect what man is—warts and all. [The operative phrase.] Try to remove the warts, fill the pores, cut the calluses, and pretty soon you have a bunch of lusterless wax moon faces."[1]

It's kind of a harsh indictment, but it deals directly with the idea of book burning and, to me, any restriction whatsoever on the free access to information. In fact, as one nice little sideline, there is a senator, was, past tense, she was not re-elected, a senator from Texas known as Paula Hawkins. Sorry?

STUDENT: Florida.

SASHA: Florida. Sorry. Many people get Florida and Texas turned around in the drug area. One of the reasons I chose the textbook *Chocolate to Morphine* is that she actually held it up on the floor of the Senate and said, "This is the reason we should ban books in the schools." There is no reason to promote educational process on the appropriate use of drugs. Her statement was, "We must just teach abstinence." That's one reason you have the book you have. Because it is a good presentation of the warts-and-all of dealing with drugs. I completely side with Dr. Weil—provide honest facts, the goods and the bads of drug use, and let the student determine their own relationship with drugs and their use.

---

1 Ray Bradbury 1953. *Fahrenheit 451*, page 79.

Okay, where am I? Definitions. What is a drug? Wait! One more thing before we can get into this. One of my own personal little hang-ups—again, I was talking about my hang-ups—is I hate the idea of people scribble, scribble, scribble, taking down notes. And then desperately trying to memorize the notes. I'm forced to give a midterm. I'm forced to give a final. I don't want to. I was told to. "You give a midterm. You give a final. You give grades unless they say they don't want grades." But what I do want you to do is get the music of what I'm saying, get the flavor of what's going on, and so I have never given a course in my entire life that is not open book. Bring in all your textbooks, bring in all your notes. In fact, what I often do is give a midterm so you can take it home and ask people the answers to questions that you don't know. Because the function is to get the answers to questions, to get the feel of what's going on. Everything is open book. Take notes if you want. What I did finally at my Berkeley class this last year was I wrote out all the lectures ahead of time, which took one hell of a discipline, but it was a big wad of paper like that [gesturing]. Everyone had the lectures of the entire course ahead of time. And so, all we had to do was come in there, "You've read the lecture? Fine. Any questions?" No questions, we go home. [Laughter from class.] Somehow there are always questions. But that is the spirit of what I want to do.

I would like to start with trying to find some definitions of terms that will be used off and on during the year.

What is a drug? The definition of a drug is probably one of the most difficult of the lot. Some will be extremely narrow in the definition, requiring that it be a chemical that is used for the treatment of some illness. Others will widen the scope to include anything and everything that in some way affects the living system. Often people will answer that with that marvelous, marvelous, catch all phrase, "Well, you know what a drug is," or "Everyone knows what a drug is."

What is a drug? I think if I were to say aspirin, here's aspirin, you've got a headache, you've got a little rheumatoid arthritis that's bothering you, you've got perhaps a bit of fever, take an aspirin. Maybe it'd open up a hole in your stomach and get a little more bleeding from your ulcer, but that's okay, at least you're over your headache. Fine. It's a drug. It influences something that goes on in the body

I would say that probably on the far-out side of not-drugs you'd get that maybe a glass of water would not be a drug. Although, it will possibly give satisfaction, maybe even pleasure, possibly quench thirst and maybe change the salt balance in the body and affect the kidney function a little bit, all these being so called drug actions. But where are you going to draw the line? X-rays? Cod liver oil? Chicken soup? Chocolate? An exciting book? A placebo? Vitamin C?

Last year in the fall course at Berkeley I handed out a questionnaire. I said, "Here are fifteen things and write 'yes' or 'no' if you think they're drugs." I tallied them and I had some people who answered everything 'yes' and some people who only answered one or two 'yes.'" Is chicken soup a drug? I know a lot of people who say, "You've got a cold, perfect thing for a cold. Take a bowl of chicken soup and go to bed." You know, fine. You're treating it. Is a placebo a drug? Who knows the word placebo? Who does not know the word placebo?

From the Latin "I please." It is the use of something we humorously say does not have a drug action with the promotion that it does have a drug action and you are going to respond to it. You have a study, a so-called double-blind study, which means you don't quite know what's going on nor does the physician nor do the patients. And you say, "Here's a bunch you're going to put on morphine. Here's a bunch you'll put on aspirin. Here's a bunch you'll put on a placebo. And we'll see how they respond to a given pain stimuli." So they put a wire in on a molar and push a switch somewhere and people go "Aah!" like that [laughter]. And some people go "Aah!" loudly and some people go "Aah!" softly. How do you determine pain in human beings? Well, you gotta push something in that hurts. This is one reason you tend to use rats and mice because they're little things, they don't hurt. Ever heard a rabbit scream? Believe me, once you have worked with a rabbit who doesn't make any sound and you suddenly approach it in a way that it knows is going to be painful and you hear a rabbit scream, it's the last time you'll work with a rabbit in a laboratory.

We collectively believe that animals are lower things that are put on earth just to be experimented with, you know, and so we kill a cat or kill a dog. At least we're not hurting things that count, like people [with irony].

However, the whole area of psychopharmacology is all dealing with people. You're dealing with something that goes in. Your sense of pain is your response to pain. We're going to spend a whole lecture on the narcotics, specifically on morphine and heroin, drugs that are used to quiet the offense of pain, but drugs that don't touch the pain itself. How many people have had a toothache and taken codeine? Okay, more than half. Have you ever noticed, those who have taken codeine with a toothache, that your toothache is still there, it just doesn't bother you anymore? It's that marvelous stepping aside.

My first experience with morphine was with a wound I had during WWII and I was going into England. I was about three days out of England on a destroyer and was below decks and we were playing cards and killing the time until we got into England. I was on morphine pretty much all the time because this was one hell of a painful thing. And I was dealing with one hand, I learned to deal with one hand, and the guy in sick bay would come by and say, "Is your thumb still hurting you?" "Yeah, probably a little bit more than it had before. Whose deal?" You know, the next thing you're dealing cards. The pain is still there. It's a beautiful, powerful tool to treat pain because the pain is there, but it doesn't bother you.

Pain is a very necessary signal to the body. If you had the body freed of all pain, you would stumble around and get into real trouble in short order. Pain is a signal to the body, "Hey, hang on. There is something wrong here. There is something out of place. There is an infection. There is something that is going to hurt you." Pain is a good awareness. You want to soften the agony of the pain, but not to remove it. If you remove the pain, you use analgesia, something that is totally anesthetic. Morphine is not one of those. Morphine and all the allies of morphine are very powerful because of their marvelous property of going into the brain and saying to the brain, going somewhere in its receptor sites and saying, "The pain is still there, but for a few hours it's not going to bother you." It's a superb drug for that purpose.

So I would say a placebo, by way of this one experiment with electric teeth, I think it was morphine and placebo that were more active than aspirin. But in that particular experiment I remember that the placebo was fully equivalent to morphine. I think placebo is a drug. (I believe it's green ones that are more effective than yellow. I get the colors turned around. Green and yellow. There was a big study made on green placebo as opposed to yellow placebo. And the greens I believe are more effective than the yellows consistently.) I mean, you go to a physician—our whole entire culture is going to the physician—"I'm sick, Doc. Give me something to help." Give me something to help. It's not "Look at me" or "Get my background" or "Tell me what you think." It's "Give me something to help." We are drug oriented unbelievably to a physician. If most people go out of a physician's office without a script in hand to go down to the drugstore and stand in line to buy $22 worth of something, they feel they've been robbed somehow. That's part of the procedure. We are pill oriented and want to hear, "Take these. Take these four times a day, once after each meal without taking a glass of water."

How many people have ever been in a hospital in Italy? Aha! Ever been treated via the suppository? They consider taking pills to be basically immoral. Everything is given by suppository. Suppository: in the anus, up into the rectum. That's how you take pills in Italy. And, by golly, if you have a sore throat, "Take this." Not, "Take this." "Take *this.*" [Gesturing.] [Laughter.]

Again, the whole philosophy of how you handle yourself. But the drug orientation is instilled in our relationship with the physician, in our relationship with illness. When you're ill, you go to the physician. You don't look into yourself. You look to someone else for help.

I have down here as examples things that are in the grey area between things that are drugs and things that are not drugs. Cod liver oil. Good heavens, anyone who has taken cod liver oil knows that it has certain physiological consequences, usually for a fairly short time. That is certainly a drug action. And yet, it's certainly a perfectly proper food, and has been used as a food for millennia. A fourteen-hour workday, the workaholic, the one who escapes all his problems, his family, his children, the whole mess at home, by knocking his tail off at work, staggers in late at night and gets up early in the morning and goes back to work. He's an escapist. I'll call that a drug-type action.

I'm making this definition quite broad because I want to get into certain things such as radioactivity. I consider radioactivity in its own way to be a drug. Not only that you take drugs and put radioactive isotopes on them so you can trace them in the body, or maybe you can go to a place in the body and cook something you want to cook. There are these uses of radioactivity as drug agents.

But also, I was reading a report just recently upon the general feeling in the northern part of Italy as an outgrowth of the Chernobyl accident in Russia. And there was a general mass dis-ease in the people. They didn't know what to do with the children. Should they leave their homes? Where do they go? Are they not walking maybe into the face of the storm by going that way? The government is being totally free of candidness in not saying beans about what the radiation level is. Well, the radiation level in the north of Italy, where they had really gotten badly hit in the agriculture, they say the radiation is only twice normal. But what they are doing is taking all of Italy, including the south of Italy where there was no radiation, and averaging it all together, which didn't equal a picture of what your risks were. Then the next day, another agency would say. "No, it's 100 times normal and don't eat any fresh vegetables." The next day it'd be, "Wash the fresh vegetables before you eat them." The next day it would be, "Don't use the water for any purposes until—" All this sort of thing.

The turmoil, the agony in these people was real and they did not know where to go. As I would not know where to go, you would not know where to go, if we had old Rancho Seco suddenly blow a few hundred curies loose up there. "And which way was the wind going?" "Well, we're not allowed to tell you which way the wind was going. We're not free to give you that information yet." "Tell me!" Consider yourself in that agony. You are undergoing a lot of turmoil inside of you. There is a case of a drug action on you, which is radiation that's a thousand miles away. So, in a sense, anything that upsets social patterns, not just the overt using of drugs in society—and we have a lot of cases of mass medication that no one talks about. But I will. Nice or not. Warts-and-all philosophy.

For example (this is typical of my rambling), this is the kind of thing that I feel should be made clear to people. How many people approve of fluoride in drinking water? Basically I do, too. Well, okay, that's about half. How many people disapprove? Okay, I'd say about a half and about a third. Some have no opinion, which is a perfectly fair thing. People will say, "Fluoride is good because it does this and that to possible cavities." Maybe it inhibits cavities. Bad because it mottles the teeth a little bit, but the good outweighs the bad. People will say, "Well, fluoride is a fraud, fluoride is this." The arguments have raged back and forth. Not one person addresses the question directly which is: Fluoride in our drinking water is mass medication. You are giving a drug to everyone without their choice. And I consider there is the evil part of fluoride. Fluoride is basically good. I think it does more good than bad. So, I'm not against fluoridation on the concept of the use of the drug. I am violently against the use of an inescapable vehicle for mass medicating with a drug. They'll say, "Oh, chlorine's been in the water for years." Chlorine treats the water. Fluorine treats the person. So fluorine, fluoride is the ion, is a drug that is in water and that is mass medication. There are other such examples, the precedent has been set, and it scares me.

Philosophy aside, what is a drug? The FDA has given a marvelous, marvelous, long legal definition[2] that goes on for four paragraphs, which says, in essence, 1) a drug is something that's recognized by the US *Pharmacopeia* or listed in the *Pharmacopeia*, the *Homeopathic Pharmacopeia*

---

2 In section 201(g) of the Federal Food, Drug, & Cosmetic Act. 21 USC 321(g).

and the *National Formulary* and any of its supplements, and 2) (including more things) is an article used in the diagnosis of disease, the cure, the mitigation (which means "softening the intensity of"), the treatment or the prevention of disease; or 3) articles other than food. The fourth paragraph adds any "articles intended for use as a component of any article specified in clauses A, B, or C of this paragraph, but does not include devices or their components, parts or accessories."

So, both medical devices and foods have been specifically excluded.

An early pharmacological definition, *Pharmacology and Therapeutics* (Musser & Shubkagel) defines it in completely functional terms: A drug is "a chemical substance that affects living protoplasm and does not act as a food; it is used in the cure, treatment or prevention of disease in man or animal; and it can alleviate suffering and pain."

Professor Samuel Irwin has given a yet simpler definition of a drug: "A drug is any chemical that modifies the function of living tissue, resulting in physiological or behavioral change."

I would make the definition looser yet, and considerably more general. Not just a chemical, but also plants, minerals, concepts, energy, just any old stuff. Not just changes in physiology or behavior, but also in attitude, concept, attention, belief, self-image, and even changes in faith and allegiance. "A drug is something that modifies the expected state of a living thing." In this guise, almost everything outside of food, sleep, and sex can classify as a drug. And I even have some reservations about all three of those examples.

Food, by the way, they also define in the FDA regulations,[3] and here they really go to the other extreme. They really cop out with a circular definition. They say food is something that's used as a food. [Laughter] What they actually say is food is also defined in the following way as: 1) articles used for food or drink in man or other animals; 2) chewing gum; and 3) articles used as components for such articles. That is, it for the definition of food. Chewing gum, by the way, is legally a food.

---

3 In section 201(f) of the Federal Food, Drug, & Cosmetic Act. 21 USC 321(g).

Talking about things that are legally one thing or another... (I don't know where I'm going to get into it in the course, I don't follow notes and so if I get into something a second time, put up with me, and if I get into something a third time towards the end of the semester, someone raise their hand and say, "Get into something new.") I want to get into one thing right now that I have not talked about (the first hour is beautiful, I know I'm not repeating any old stories) and that is the concept of food additives.

I don't even know if I'm going to get into food additives. I've gotta get into it somewhere. But there is an amendment that was passed in the 1960s known as the Delaney Amendment. Those people who have been in and around public health and nutrition will be aware of it. It's an amendment that says, thou shalt not add any additive to food at any level if that additive causes cancer. And here they say: causes cancer in any animal in any way.

We're going to get into words, we might as well get these kinds of words in there. [Writing on board.] The word carcinogenic. Let me write two other words up here while I have it because they're often confused. Mutagenic. This is a lecture on vocabulary. And teratogenic. All these things have the feeling of messing up cells and getting in the way of things. I'm going to take these three words separately.

The suffix "genesis." First book of the Bible. Creation. The idea of creating something.

Carcinogenic is something that creates cancer. Carcinogenicity means, literally, the ability to cause cancer. The suffix "genicity" is found in all three of these terms, and is often seen as genesis. It refers to the origin, or "genesis" of a thing. There has been great concern about the exposure of living things to agents that can produce cancer.

First, on a complete aside, you would be surprised how awfully difficult it is to create cancer in a test animal or in human beings by administering a carcinogenic substance. I did an experiment way back in my graduate days in which I wanted to develop cancers in rats because we had a whole series of compounds that appeared to be very effective in aborting cancer development. We used butter yellow, something that's not been used as a food additive for years because it's one of the most intensely carcinogenic things known. You administered butter yellow to the rat in the morning; you administered in the afternoon; the next morning. Then someone comes in, takes the afternoon session. For the entire life span of the rat. For ten months we administered butter yellow. That rat was buttered yellow. Its white fur was yellow. Its eyes were yellow. Rats don't vomit, but its excrement was yellow. It was yellow. It had no cancer. Finally, at a year and a quarter we got two out of ten rats to get cancer. We couldn't even run the experiment to try to keep the animals from getting cancer because we couldn't get the animals to get cancer on the most intense carcinogen we knew at that time.

Now they have methylcholanthrene and other goodies that are more effective, but making an animal develop cancer is a difficult task from some of these carcinogens. In experimental animals, to generate cancer by exposures to even the most potent carcinogens usually requires repeated exposures. But it can be done.

The Delaney Amendment[4] said, thou shalt not add any food additive that develops cancer at any level in any animal under any test to food. Well, then you get into a neat thing. How many people have heard of saccharine? Aha! Good.

Talk about drugs. Is saccharine a drug? How many people think saccharine's a drug? How many people think it's not a drug? More drug than not. Interestingly, it is legally a drug for reasons that are rather humorous and I can be a little cynical about them. Saccharine actually does cause a change in body behavior patterns. Sure, it tastes sweet. It's excreted, in the urine not feces, largely unchanged. (I'm going to get a little bit into, I think in the fourth lecture, what I call the body's plumbing, how things go through and how they come out.) But saccharine goes into the mouth and it's excreted in the urine about 97, 98 percent unchanged. So it largely goes through the body, but it goes into the body because the

---

4 Delaney Amendment, which amended the 1938 Federal Food, Drug, and Cosmetic Act. It was enacted on September 6, 1958, as Public Law 85-929.

tube through the body is mouth and anus. To get out the urine, it has to be accepted into the body and then be secreted separately into the urine. So it goes into the body; it doesn't do anything. It doesn't have any metabolism. It has been shown to have no pharmacology. It is strictly something that once it's in the mouth tastes sweet often with a little— anybody tasted pure saccharine as a chemical? You have that metallic-y yuck, as if you've been sucking on a piece of aluminum foil. Which is not really the nicest kind of a feel, but there's that slight metallic aftertaste. Now they blend it with other sugar substitutes.

If you measure a person's sugar level, the insulin-controlled sugar level in the body, and you take a little sugar, the sugar goes into the system, the insulin goes up, regulates the blood sugar, it comes back down, the cells accept the sugar in the presence of insulin. Your body sugar level does pretty consistent things thanks to the ins and outs of insulin, which is the hormone that controls it. And you take a little bit of saccharine, the signal goes to the brain, "Hey, something sweet's going in." The brain does its thing. Out comes the insulin. The sugar level drops because you're taking in something that doesn't have any sugar value. And so, suddenly the sugar level drops, the insulin pulls back and stops secreting, it goes up again and you get what I call an Oscar Levant syndrome in which you overshoot, undershoot, overshoot, undershoot, and dampen out. The fact you've started blood sugar oscillating by using a drug that has no pharmacological effects means it has a pharmacological effect. That has nothing to do with the cynicism that comes with it as a food additive. The carcinogen may sometimes be balanced on a fine line separating drugs from not-drugs.

It has been found, in one study—a lot of people say it's really not a good study. You'll find the phrase "that's not good science" used time and time again when someone runs an experiment and comes up with a conclusion that you disagree with. You'll say "that wasn't good science, it wasn't done in a double-blind system, it was a preliminary study that doesn't have any value." Marvelous, marvelous euphemisms for throwing out other people's results that don't agree with your philosophies. I don't think I'll get into anywhere else in the course, so I'll do that now—anyway, a study was done in Canada in which they gave saccharine to a bunch of mice and they developed bladder cancer. Not all. Two out of ten. Some developed bladder cancer from saccharine. The fact that they put saccharine in a little capsule and implanted the capsule in the bladder may have been more causal to its actually developing that kind of cancer, but that was not the argument. The Delaney Amendment said any chemical that produces cancer at any level cannot be used as a food additive.

The saccharine manufacturers screamed, "You can't take that off the market. Many people depend upon it. People who are overweight. The entire obesity crowd drinks saccharine all the time. You've got to leave it on the market." Well, big battles and, of course, you kind of know how it came out. Saccharine is still available as a sweetening agent in a lot of soft drinks. How much is this? A hundred and twenty milligrams of saccharine. A lot of stuff. Okay. But it can't be in there as a food additive, so the FDA says, "We have hereby declared that saccharine is a drug." And there's no law that says you can't add drugs to food if they're carcinogenic. You only can't add food additives to food if they're carcinogenic. So now saccharine has been classified as a drug. It can be used in food. It's still carcinogenic, widely sold, and there are no laws being violated. So there are very nice ways of handling it. Talk about drug abuse? Wow! There's an example. But that's the way it is, and be aware of the fact that saccharine is still used in spite of the fact that it's carcinogenic.

There is now a term that's come into the FDA known as *de minimis* which is, in essence, a cute Latin way of hiding the fact that if it is so small an amount that it wouldn't cause cancer to any appreciable effect it's okay. Part of the reason is that "in any amount" has become a real tricky phrase because every time you turn around there is now a GC-MS with ten to the minus ninth capabilities of getting pico—what could be smaller than pico? *Nanofemtoattogram* levels of drugs? That you can always find a trace of something. Yes? [Directed to Ann.]

ANN: What is GC-MS?

SASHA: Gas Chromatograph-Mass Spectrometer. It's a big instrument that costs $100,000 and picks up trace amounts of things. The thing is, if you look closely enough, you're going to find what you're looking for. And, a little sidelight, I think I'm going to talk about the history of the law in a couple of lectures, but a point comes in right now that's a good one. We just had a law passed that says the possession of cocaine as a freebase in any amount is a mandatory five years in prison without the possibility of parole. The detection of freebase cocaine in any amount. How the hell little is any amount? The detection of ecgonine, which is one of the cocaine breakdown products, in any amount—for the second time, I think it is, or maybe it's in a larger quantity, of which in there can be any amount—not how much cocaine is there. It's how much freebase is in the cocaine. They're so scared about this whole crack thing. They came out big business until the elections. Now you haven't heard about it; of course, the elections are over.

Anyway, the whole thing that promoted this bill getting passed—you will buy a bottle of Coca Cola, make a dichlor extract, make a basic, neutralize the carbonate, make a dichlor extract—don't worry about these terms. I'm appealing to the chemists who will nod and smile—make a dichlor extract of that Coca Cola, put it down to a fine thing, and stick it in a GCMS. *Bwoom!* There's the peak for ecgonine. And you turn the gain up a little bit, about one Coke bottle in five will peak for cocaine. Holding a bottle of Coca Cola in your hand is possession of ecgonine to some extent as a freebase, and in many cases cocaine to some extent as a freebase. And if a person wished to pursue that law and push it because he didn't like the way you part your hair, you're five years in prison without the possibility of parole. It's a ridiculous law, but it is the law. It will be selectively enforced—and I want to talk a lot about the laws in the third lecture, about why they have been put in there and how they are ignored, or how they are enforced when it is appropriate to enforce them.

But that's the kind of law I'm very disturbed with because they don't say how much is a detectable amount. Any detectable amount. You turn the gain up with some instruments, you'll always get a noise level. And if that noise level falls in the slot you're looking at, you have a detectable amount. It may have nothing to do with reality, but it's the number you've put on a piece of paper.

Beware of some of the instruments. We're going to get into this when I talk about urine testing. The presence of marijuana in any amount should constitute reason for further testing—they don't look for marijuana, they look for specific things that react with an antibody that has been made in response to a material that comes from marijuana. The first testing had no justification and to test further on the basis of something that could be present at the noise level is irresponsible. Therefore, you put a cut off up here and anything below that we're going to call negative.

You have cut offs in almost all these tests where you have very variable degrees of sensitivity. But if you wish to nail someone—for example, two guys were on a locomotive that went through a stop sign and went through something or other and went into a passenger car killing fifteen people on January 4. I saw in the paper, it's Amtrak's biggest catastrophe ever. Apparently, they found evidence of marijuana usage. What is meant by evidence of marijuana usage? If you were to come in this room and there was a cloud of marijuana smoke because there had been a wild party in here during someone's birthday the hour before, and you say, "I don't like the smell of this place" and you clear out of there, you are going to be marijuana positive for a few days. Just from having walked into and out of that room. There is no evidence that you've used marijuana. In fact, when you get down to that noise level, there's no evidence that what you're looking at is marijuana. But we've got to have a drug to nail our projections on.

You have a guy who's in the top of a campanile in the University of Texas and he's just nailed fifteen nurses, and finally some sharpshooter gets him and he falls out of the campanile, and he's on the ground and they haul him off to the morgue to run a test. The desperate thing is find a drug. Find a drug! First thing that goes out in the newspaper: The madman marksman who shot fifteen people was finally killed—they suspect drug use. How do you know when you're carrying a guy in the stretcher off to the morgue that you suspect drug use? You're desperately hoping you'll find a drug, because if you find a drug, then you can say, "PCP has claimed another victim!" You know, "The evil drug" PCP's our current evil drug. Anything we don't want to realize about ourselves we now load on PCP. For a while it was heroin. Before that it was marijuana. God knows what it's going to be next week. But right now it's PCP. And crack, by the way. Everything that's bad is nailed on crack. Okay.

You suspect he's under some drug effect because you say, "No person in his right mind would shoot fifteen people from the top of a campanile at the University of Texas." I don't know if that's where it was, but somewhere down there. I think a lot of people would like to shoot fifteen people from the top of a campanile in Texas! [Laughter.] "Boy, I know I can get fifteen more before they get me. All the better!" You can see where that comes from, you know—We've all got that beautiful, beautiful corner of our mind, you know, "Screw them! I'm gonna get them before they get [me]!" It's in there! You choose not to pick up a rifle and go out and shoot fifteen people. But if something slips its little category and you slip into that marvelously pleasant and maddening psychotic state in which you are not at all averse to shooting fifteen people, it could be you, it could be me, it could be him. The thing is we don't want to acknowledge that it could be me. So, if it was him, it's because he was on drugs, and we're going to blame the drugs. Well, sometimes it works, but often, often it does not.

You have the same thing in alcohol. Someone is in an auto accident, alcohol is suspected. And indeed, alcohol has been involved in a great number of the automobile tragedies. But you try to find out how many people are on the highway that are not involved in automobile tragedies who have alcohol in them. You find out, if you're walking down the street, on Post Street at ten o'clock at night, a lot of people have alcohol in them. Alcohol is pervasive. In fact, one whole lecture's going to be on alcohol. It's one of my four major drugs. Alcohol is one of those. You find people are all the time into it. And, of course, if they're in an accident and they're into alcohol, you say the alcohol was involved in the accident. And indeed, it was, in a sense. But if they're not in an accident, alcohol's involved not in an accident, too. So you don't want to ask what percentage of people who are in accidents have been involved in alcohol, but how does the percentage of people in accidents involved in alcohol compare with the people as a control group who are not in accidents. And you'll find the numbers are not as radical as they appear to be.

There are admittedly absolutely irresponsible drunken situations in driving. But you'll find that people carry around a great collective inventory of alcohol on board and manage to find ways of behaving quite normally and handling themselves. Ever see a person try to inhale while talking to a policeman who's pulled them over. It's been done. I rode shotgun on a highway patrol one time in conjunction with an alcohol study in which I actually saw the field sobriety test being given and saw the behavior of people. People will come out of the car totally unaware of the fact that their feet are not working right, the fact that when they pulled their wallet out it went on the ground, and when they reached down to get it, they couldn't get up again. I mean, all this behavior, they're blind to it. And when you see that, it's quite conspicuous. And you see how many people ahead of you on the road, on the bridge, at midnight are weaving all over the place. They're there and God is with us. Most people survive it and get through it. There are horrendous tragedies that come from it. But the comparison is not made, blame the drug/look at the person. Their relationship with alcohol is not good. As I said on the whole issue of smoking: Their relationship with tobacco's not good.

If you can drink modestly, if you can use tobacco modestly and have a choice, have freedom of choice, and choose to do it and you have a good relationship with it, and it applies to alcohol, it applies to tobacco, it applies to LSD, it applies to heroin—there is nothing intrinsically evil about any of those drugs. Drugs are not intrinsically evil. In fact, we are going to get into the question of what is drug abuse. The problems that are bothersome with the definition of the word "drug" are nothing compared with the ones that are to be faced with the word "abuse."

What is drug abuse? In the context of drugs, the dictionary limits the definition to a single example: Improper use or treatment, and the specific synonym given for "drug abuse" is misuse. And misuse is in turn defined as incorrect or improper use. The implication is that there is a right way, and a wrong way, to use a drug. This says nothing about the drug itself, nothing about the reason for its use, but simply speaks to the way the drug is used.

I have collected over some time a number of examples of the use of the term "drug abuse," or now, more broadly, "substance abuse" as found in articles, essays, or public talks. There are four entirely separate classes for the definition of the term "abuse," and they can be collected around the four operative words *what, who, where* and *how*.

People often argue the *what*. What is the drug that's abused? I don't like it. This is the only category that assigns the term "abuse" directly to the drug itself. I'm going to start, with the little time I have left, writing some of these things down. FDA: Food and Drug Administration. I'll talk more about it when I get into the law, but it's the body of the government that originally was geared to see that drugs were correctly labelled, honestly labelled. Then that they were honestly labelled and are also efficacious. And then that they were honestly labelled, efficacious, and not in a position to be abused. Not in a position to be abused eventually meant a prescription. All this came about decade by decade by decade from the FDA, originally in the Department of Agriculture. Now in the Department of Health.

But the argument of the FDA is, "If we have approved a drug, or we have exempted a drug from needing approval, its use is proper. If we have not approved a drug, its use is abuse." They define abuse as whether they have not sanctioned a drug; in which case any use is "drug abuse." Look it up in the book. If it's in this column, it's not abusable. In this column, it is abusable. You get this category of assigning abuse to a drug, to a white solid.

Who here has heard of Gene Schoenfeld, Dr. Hippocrates? That may be another generation. Okay. A few from another generation. He always would start his lecture on drug abuse with a slide and the slide would have a pill on an anvil and there was a great big hammer above it. And you could tell by the streaks going behind the hammer that the hammer was descending on the pill. In a moment it's going to hit that pill and just smash it to smithereens. He would say, "There is a true example of drug abuse." [Laughter.] That was his argument.

The powder, the pill, the solid, does not have abuse. You speak of, "This drug is dangerous and that drug is not dangerous." All drugs have the potential of danger. All can be used *without* danger. Depends on the amount and the relationship and the person's reasons. But you now have the DEA, the Drug Enforcement Administration. Before, it was the BNDD (Bureau of Narcotics and Dangerous Drugs). Before that it was the BN (Bureau of Narcotics). And before that there was nothing. This is a progression. We'll talk about this under the law. This is the progression that we've gone through on the agencies that regulate drug laws. From nothing to the Bureau of Narcotics (BN) to the Bureau of Narcotics and Dangerous Drugs (BNDD) to the Drug Enforcement Administration (DEA). So this is the progression of the drug enforcement body.

But, the BNDD, the Bureau of Narcotics and Dangerous Drugs: "If we regulate them, they're dangerous. If we don't, they're not dangerous." Who's read the term "hard and soft narcotics?" Hard and soft. What the hell is hard and soft about a narcotic? I mean, if you scrunch it in your hands or it grits between the teeth, it's hard? And if it flows, it's soft? I frankly do not know what a person means when they say they were hooked on hard narcotics. If anyone can give me a rationalization for hardness or softness—

This concept of good and of bad drugs has many cousins in pharmacology. Some people classify drugs as toxic or nontoxic, or as dangerous or non-dangerous. "These drugs are toxic." "Those are not toxic." This assignment of an intrinsic *abuseness*, or toxicity, or dangerousness, to some drugs and not to others is nonsense. All drugs are abusable, all drugs are toxic, and all are dangerous. And all can be used sensibly and safely, without toxic effects, and in a way that is free from danger. All drugs in small enough amounts are nontoxic. The FDA classification may be simple, but it is useless. [Directed to student] Yes?

STUDENT. What's the word toxic?

SASHA: Toxic. To produce toxic symptoms. I copped out by using the word as a definition. Symptoms that give you evidence of a life-threatening situation, such as twitching, convulsing, vomiting, unconsciousness, something that clearly, conspicuously indicates there is something wrong with the organism. As opposed to lethal, which is something that actually kills. Things that are toxic can be lethal and things that are lethal invariably go through a toxic phase. But in essence, the argument is nontoxic/toxic. Clearly, if this amount of drug will kill, that is lethal, something less than that is toxic, something less than *that* is nontoxic. Every drug you wish to name has each of these categories.

ANN: The same meaning as poisonous?

SASHA: Poisonous is a general term. It means it produces some untoward effect. But these are really shades of the same thing. In fact, I was going to get into the whole area of carcinogenic, which means the genesis of cancer. Poisonous comes from this. Mutagenic, the genesis of mutation. And teratogenic, *terato* meaning monster, the generation of monsters. Let me wind up on these three terms because they are often kind of mixed up in the mind.

If you give a drug to a test organism, or to a structure of some kind, and the structure develops cancer—truly carcinogenic. Most things cannot be shown that way, so instead you look for mutations. This is a process that is known as the Ames test. Ames, from a professor of biochemistry at Berkeley who developed a mutagenic test in which you took an organism that would not grow because it lacked a certain fundamental thing for growing. It came from an organism that would grow because, let's say, there is an amino acid that it needed for growth and it had to obtain it from its food. This is a mutant of this organism that won't grow without that. So, what had been not needed is suddenly needed.

You take this organism, you sit it in a petri dish in a warm place and nothing grows. Oh, a few of them spontaneously change their genetics and start growing, but you have very small growth levels. Then you sprinkle a drug on it, or sprinkle a person's urine, or you treat it with something or other, and all of sudden a lot of these organisms start growing. The extent of growth is a measure of the potency of the chemical as a mutagen. They have mutated. So that drug or urine contains something that causes these organisms to mutate. And mutation means they have changed their genetics and it's kind of close, in a sense, to carcinogenic, where a cell will divide without regulation. They are not the same. Things can be mutagenic that are not carcinogenic. And things can be carcinogenic that are not mutagenic. But the track record is good that most things that are carcinogenic are also mutagenic, so now automatically this test can be run easily. This is very hard. So you run mutagenicity tests on foods, on people's urine, on biological systems, to see if they have the capacity of affecting the genetics of something.

Teratogenic is completely different, very straightforwardly the creation of monsters. This is the property of a drug that when it goes into a pregnant woman at a period of sensitivity, usually in the second month or thereabouts, will affect the development and the laying down of the embryo. And it comes out with some malformed embryo as a consequence. There is a period of sensitivity during pregnancy, in humans somewhere within the first several weeks, when the developing child is exquisitely sensitive to exposure to foreign chemicals. A general rule should be inflexibly observed: Use no strange chemicals (foods or drugs) during pregnancy, especially during the first trimester (first three months).

The most famous example of this was the sedative-hypnotic thalidomide, back about twenty years ago, in which thalidomide which was never approved in this country but was widely used and distributed, gave rise to a condition of embryonic development known a phocomelia. It's where the child is born with flippers (each called a *phocomelus*—a name derived from a seal's arm) instead of arms. It's a tragic situation. There were many of these things—that is because of a drug given at the time of sensitivity of the embryo. The concept is known as teratogenicity and the drug responsible is teratogenic. It forms monsters *in utero*.

A second common definition of drug abuse is the use of drugs outside of a physician's purview. Or *who* dictates their use? If an MD says "Use a drug for this or that purpose," then that is drug use, and if one self-medicates, then that is drug abuse. A quotation of Dr. Thomas Szasz, in his book *Ceremonial Chemistry*:

> "Society's prevailing view is that being medicated by a doctor is drug use, while self-medication is drug abuse. This justification rests on the principle of professionalism, not on pharmacology. [This] concept of drug abuse symbolizes scientific medicine's fundamental policy that laymen should place their care under the supervision of a physician. This is similar to the belief, prior to the Reformation, that laymen should not communicate directly with God but should place their spiritual care under the supervision of a duly accredited priest. The self-interest of the church and of medicine in such policies are obvious. These policies also relieve individuals of the burden of responsibility for themselves."

There is one form of drug administration that is defined by the "who" involved, which is, in my book, inarguably the abuse of drugs. This is the administration of a drug to another, where that second person is in ignorance of the act. With adults, there have been an increasing number of criminal actions such as robbery that have seen the surreptitious giving of drugs to incapacitate the victim. Common examples in recent reports from New York have been scopolamine (a parasympatholytic, leading to a state of confusion, delusion, and active hallucinating) or lorazepam (a benzodiazepine which is a sedative and leads to confusion and possible coma). A separate form of drug abuse along these lines is exposure of children *in utero* to drugs during pregnancy.

If you are pregnant or you think you're pregnant, and you are the perfect sex to be so, don't use drugs in that first trimester at all. All drugs carry the potential of teratogenicity. In fact, stay away from alcohol and coffee. You do not test in human animals whether a drug is or is not teratogenic—don't take the chance. But that's my own strong opinion.

Our third category of drug abuse—*where* are they obtained?

The illegalness, or illicitness, of the source as well as the paramedical aspect of drug use is "abuse" in some authorities' eyes. Dr. Jerome Levine, at the NIMH, has defined drug abuse as:

> "The self-administration of [drugs] by individuals who have procured or obtained them through illicit channels, and/or in medically unsupervised or socially unsanctioned settings."

The clear feeling here is that if you interact with drugs against the law, or against the accepted social philosophy, you are abusing the drugs. This, in the final analysis, makes the legal structure, or the immediately accepted dictator of social mores, the source of the definition of drug abuse. And since the laws change continuously, and the social structure can vary so completely from one area to another, or from one culture to another, this definition is useless in addressing the problem.

And finally, *how* are they used?

I personally believe, most strongly, that in the improper use of drugs lies their abuse. Dr. Irwin has phrased it thusly:

> "[Drug abuse is] the taking of drugs under circumstances, and at dosages that significantly increase their hazard potential, whether or not used therapeutically, legally, or as prescribed by a physician."

The medical community largely agrees. The Advisory Panel on Alcoholism and Drug Abuse, of the AMA Council on Scientific Affairs, defined the following terms:

> "Use—the taking of drugs pursuant to proper medical indications; Misuse—the taking of drugs for non-medical indications; and, Abuse—drug taking that interferes with a person's health, or economic or social functioning."

But even this is a little self-serving, in that the pejorative term "misuse" is still the blandest category that functions outside the control of a physician.

People use drugs, have always used drugs, and will forever use drugs, whether there are physicians or not. I prefer the ideas of Andy Weil, who holds that drug abuse has nothing to do either with the legal or medical approval or disapproval of the drugs involved, or with the reasons for their use. Any use of a drug that impairs physical or mental health, that interferes with one's social functioning or productivity is drug abuse. And the corollary is also true. The use of a drug that does not impair physical or mental health or interfere with social functioning or productivity is not drug abuse. And the question of its illegality is completely beside the matter.

Three terms have been used from time to time to describe the pattern shown by everyone at one time or another concerning the repeated use of drugs.

Habituation: This is a reasonably benign term, referring to a repeated drug use without serious harmful effects to either the individual or the society. In a sense it can be compared with the "sweet tooth" of the person who craves candy.

Addiction: The term "addiction" has fallen out of usage in the current medical literature, but it is still popular in the lay press to condemn drug users and especially certain drugs. It is a psychologically loaded term, implying compulsive abuse associated with physical dependence. There are negative effects implied to both the individual and the society.

Dependence: This is the term that has now come into general use to cover both the psychological and the physical aspects of repeated drug use. A careful and useful essay is Dr. Maurice Seevers' in the *Chemical and Biological Aspects of Drug Dependence*. This has been of great use in framing this concept. A number of alternate meanings of dependence must be considered separate from drug dependence.

There are many habits, routine behavior patterns, that need not be associated, but which in their own way rob us of some aspect of freedom of choice. These are actions that verge on the involuntary, but upon which we have come to depend for our "normal" daily life. And there are drugs which can be used for replacement therapy and upon which our continued health may well depend. This type of dependency (to things such as vitamins, insulin, chronic medication) must be excluded from consideration here. And there are uses of the concept in areas such as microbiology, where one can see an adaptation of bacteria to drugs such as inhibitors or toxins, but this is a matter more of genetics than of psychological compulsions.

A fair working definition of dependence is the establishment of a conditioned pattern of drug-seeking behavior through the repeated use of the drug. The term is usually used in connection with psychoactive drugs, and has been broken into two subdivisions, psychological dependence and physical dependence.

Psychological dependence is the branch wherein the drug user finds that their psychic response to the drug is sufficiently attractive to motivate a repeat of the experience. In physical dependence, on the other hand, there occur neural or metabolic changes in the body that result in an improved accommodation to the drug, as seen by either (or both) tolerance or withdrawal reaction.

Relapse: To relapse (in the area of drug jargon) is to return to some earlier pattern of drug relationship that had been agreed upon (personally or socially) to be wrong. A "slip" is a single or perhaps a few cases of some drug re-use, and in many philosophies is not considered a failure in the noble effort of trying to change one's behavior patterns.

The area of evaluation here embraces not only drug use (in the social sense), but sexual behavior (check both the confessional box at the local chapel and the political scene) and one's interaction with alcohol and tobacco (if one is intimately involved with these particular problems). The reasons for relapse may be biological, but there is a lot of rationalization that goes on in the mind.

Craving: This is the intense desire that becomes an inescapable obsession for a drug (or sex or politics or food or…) that an ex-user comes to feel for the thing that they have brought under control. It is intense early in the separation, but it never completely disappears. Ask an ex-smoker, after they have been without cigarettes for some twenty years, "Do you miss them?" and they will probably say, "You're damned right I do." This is a lifetime burden that is part of the withdrawal package from an addictive drug.

Tolerance: Tolerance is a decreased response within a person to a given dosage of a drug, following an earlier exposure to that drug. This declining effect often prompts the increase in the self-administered dosage. It can develop rapidly and has been known, with some drugs and with certain people, to be observable after only a single exposure to a drug. In this case, it can be called acute tolerance or tachyphylaxis. When tolerance has developed over a period of time, following many exposures, it is called chronic tolerance. With some drugs, the increasing of the dose administered (needed to produce the desired effects) will reveal some new pharmacological properties that were unsuspected at normal doses. The production of a psychotic state following high uses of Amphetamine by the tolerant individual is a well-studied example.

An aside is useful here to bring in the definitions of some of these additional words, since this is, after all, a section devoted to definitions. Acute has a popular meaning implying severe or extreme. But in medicine the usual usage is to emphasize one-time, or a single occurrence rather than recurring, although there may well be a rapid onset of the indicators of the condition under consideration. And the companion term chronic means that something has been repeated, or experienced many times, again and again. There is an intermediate term meaning not just once, but then, not many times either—just a few times. This term is subacute. Another word above was the term tachyphylaxis. "Tachy-" is a prefix meaning fast, and "-phylaxis" derives from the Greek word for shield or guard. This suffix is found in other terms such as prophylaxis (a guard against, as in protection against disease) and anaphylaxis (a guard upwards or backwards, representing the medical concept of sensitization).

Metabolic tolerance (or dispositional tolerance) is a loss of responsiveness due to actual changes in the body's capacity to metabolize or biotransform the drug. This usually follows the stimulation of enzyme systems in the liver that are responsible for inactivating the drug in question. Pharmacodynamic tolerance is a term implying the adaptation of the body (the receptor sites or actual areas of drug action) in a way that decreases the intensity of the response.

Cross-tolerance is the recognition by the body of a new drug following the development of tolerance to an initial different drug. The two drugs are usually related to each other pharmacologically, and quite often share some structural features.

Withdrawal: When a person abruptly stops using a drug, there is developed a state of withdrawal which is also widely referred to as abstinence syndrome. The term syndrome refers to a collection of symptoms that has been used to define a medical illness or a physical state.

The most common type of dependence (psychological) leads to displays of emotion of exquisite rationalization (for continued use of the drug that has been withdrawn), and often to some form of alternative pleasure seeking. However, in examples where drug usage has led to physical dependence (physical addiction), withdrawal can be modest or it can be life threatening, largely dependent on the drug and on the history of use (time and amount).

In the instance of morphine and its related agonists (an agonist is a drug that produces similar effects, whereas an antagonist is a drug that blocks the expected effects) the most common physical signs are abdominal pain, diarrhea, nausea, and vomiting. The abstinence syndrome associated with heroin withdrawal has been compared with a very bad cold. This syndrome will occur in a day or so following drug cessation, or virtually immediately following the administration of an opiate antagonist.

With the barbiturates, the syndrome can be dangerous. One sees effects that resemble those that are characteristic of a convulsant drug. Within the first day, there is neural hyperexcitability and tremors. In the second to the eighth day, there can be grand mal convulsions with both tonic and clonic character, and mental blackouts. There is commonly a delirium and a psychosis-like state with thought disorder, hallucinations, and paranoid delusions. The generalized weakness can last a couple of months and there is usually a REM rebound.

With alcohol, this withdrawal syndrome can be more modest (following the acute exposure, the hangover, or simply sleep disturbance and irritability) or it can be expressed as a full-blown delirium tremens (everything seen with the narcotic withdrawal, including convulsions, which are more often of the *petit mal* type).

More vocabulary. The term hyperexcitability means more excitable than usual, and introduces a neat set of prefixes. When something is normal, the appropriate prefix is "eu." Thus one sees terms such as euthyroid, meaning that the thyroid is acting as it should. (An interesting viewing of our present acceptance of the life around us is the term euphoria which by its origin means "normal feeling," but in our culture has been equated to the excited hyper-state that we seek out and try to escape into.) The opposite prefix is "dys," which means not normal. Thus, an abnormal thyroid is called dysthyroid, but it can be too high in activity (hyperthyroid, from "hyper" meaning above) or it can be too low in activity (hypothyroid, from "hypo" meaning below).

The terms tonic and clonic are often intimidating when seen in reference to convulsions. A simple mnemonic trick can keep them straight. With tonic, remember tonus, which is a term referring to the firmness (the tone) of a muscle. Thus, a tonic convulsion is one in which the muscle becomes rigid and distended, completely inflexible. With clonic, remember a cyclone in which the weather bashes everything back and forth. A clonic convulsion displays in alternation of contraction and relaxation of the muscle, providing a sort of thrashing action.

REM stands for rapid eye movement, a phenomenon involving the coordinated searching with the eyes right and left, up and down, that occurs during sleep and which is felt to be associated with dreaming. With certain drugs such as the barbiturates, there is a deficit of REM (and thus presumably of dream time) and it is now believed that a certain amount of dreaming is necessary for the mind and body to be in good health. Then, with the withdrawal of a person from dependence on barbiturate drugs, called REM rebound, there appears to be intense and not necessarily pleasant dreaming.

And the term mnemonic refers to a device (or a trick or a strategy) for helping one to memorize a word, or a series of names. As an example, the three Baltic states, in north to south order, are alphabetical: Estonia, Latvia, and Lithuania. From the same source is the word amnesia, meaning without memory.

Lethal, toxic, poisonous—these are three more occasionally interchanged words that were mentioned earlier. Lethal is the most straightforward. This is the level of a drug that will kill a living thing. A common, and I feel largely useless, pharmacological procedure is the determination of the lethal level of an experimental drug. This is the measurement of the LD-50 of a material, a term that means the dosage level that will kill half of the animals to which the compound is given. Alternate terms such as an LD-95, or an LD-5, have been used, and these represent the lethality of a given dosage to 95 percent, or to 5 percent, of the test animals. The term is used to predict the potential lethality of the drug to humans, perhaps only by comparing a series of compounds to guess as to the least threatening.

A related measure is the effective dose, the ED-50, where half of the test animals will show some desired effect. When one looks for, say, a sleeping aid, one can measure the recovery of the righting reflex in a test animal to determine the sedative-hypnotic potential of a drug, but when the effects sought are of a psychopharmacological nature, animal tests of effectiveness are indirect at best. The quotient of the two values is called the therapeutic index (TI), and it is obvious that a drug manufacturer would like the value to be as high as possible, if for no other reason than to provide a safety margin in cases of overdose. A value of 1,000 is wanted (the lethal level 1,000 times the effective level), one hundred may be tolerated, and yet some drugs in common use such as alcohol have a TI of only about five.

The term toxic refers to any undesirable effects that may be produced instead of a desired response or along with the desired response. A material that produces only undesired effects is called a poison.

Here is another collection of words that are easily used interchangeably, and which mean many things to many people. There is no standard definition, so the best I can do is to describe how I use the terms: hallucination, illusion, delusion, fantasy, and imagery.

Hallucination is simply the seeing of something that others believe not to be there. The term hallucinogen is easily used, both in legal language and in the popular press, to describe any and all compounds that provoke (or allow) changes in the visual field. And many of these so-called hallucinogens do indeed lead to perceptual alteration, but very few of them inspire the seeing of things that are created *de novo* (with no point of origin). Most visual syntheses are distortions rather than creations. True drug-induced hallucinations are rare.

There can be a subclassification of these visual distortions, retinal as opposed to associative distortions. A retinal distortion involves the fine detail of the *macula lutea*, the small yellowish spot to the side of the center of the retina where there is the maximum visual acuity. Look at a fine, straight line. There is no exact straight-line arrangement of cells in this high resolution area of the eye, so what occurs is that the image of the line activates the cells nearest to the image, some to one side and some to the other. The other eye does its things similarly. It is in the associative area of the visual process (near the back of the head) where there is the choosing of which eye to respect and what data to use, since somewhere you know that the target object should be seen as a straight line. But with some drugs that cause distortion in this detail, the choices are different, as the intended target is not clear, or comes from some unconscious source. Elaborate that simple-line concept to a complex image (as with the surface of the running wax on a candle) and strange shapes can be seen. Such an image rarely can be maintained for more than a fleeting moment.

The associative visual distortion is closer to the process better called an illusion. There is an interpretation that comes from the visual process, in assignment of identity or meaning to something seen, that is inspired by some aspect of light or color or motion, and which is confirmed by some reinforcement from your own imagination in response to it. Sinister faces may be seen in the ivy leaves, or a particular flower may glow and call you to it.

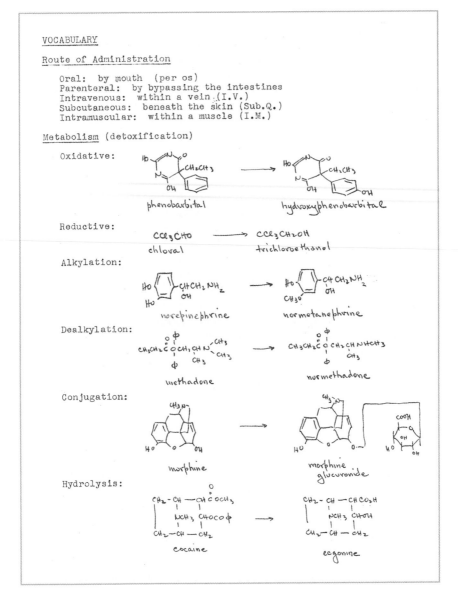

VOCABULARY

Route of Administration

Oral:　by mouth　(per os)
Parenteral:　by bypassing the intestines
Intravenous:　within a vein (I.V.)
Subcutaneous:　beneath the skin (Sub.Q.)
Intramuscular:　within a muscle (I.M.)

Metabolism (detoxification)

Oxidative:

phenobarbital　　　　　hydroxyphenobarbital

Reductive:

$CCl_3CHO \longrightarrow CCl_3CH_2OH$
chloral　　　　trichloroethanol

Alkylation:

norepinephrine　　　normetanephrine

Dealkylation:

methadone　　　　　normethadone

Conjugation:

morphine　　　　morphine glucuronide

Hydrolysis:

cocaine　　　　　ecgonine

Auditory distortions are more commonly seen in spontaneous mental illness, and are correspondingly rare in drug experiences. The hearing of voices (I was told to do this, to go there) is the province of the psychotic. More often associated with the drugs is the amplification of sound, the selective hearing of one component of a mixture of inputs, and occasionally a distortion of pitch or of timbre of the voice or a musical note. Modifications in the other sensory inputs, such as the sense of smell, are extremely rare.

A delusion is the interpretive state that can follow from a combination of the above distortions. It is an impression that instills itself as a belief that has no focus that can be confirmed. No one else can truthfully document it, yet it can haunt you with a feeling that you unshakably know it to be true.

Most of the above distortions are associated with the visual process. With the eyes closed, most of these phenomena cannot exist. The last two terms above, fantasy and imagery, are part of the human experience that is separate from the visual process.

Fantasy is the "where you are" of the imagination. You can construct a dreamlike world in which you participate, with associative interactions and imagined voices and imagined contact. All senses, except for the visual process, can be involved in this created reality. Imagery is that "what you see" companion in this eyes-closed world. Here one can construct shapes, castles, arabesques, and assorted visualizations, often with their origins in sound inputs such as music. Both worlds can occur simultaneously, with an effective closing off of the reality about you.

Combinations of these mental phenomena can be part of a drug experience, and they can occur spontaneously. No one is a stranger to these worlds, as they occur to every person daily in the state known as sleep.

This is an introductory outline of some of the vocabulary that will be used, and added to, during the course of these lectures. I will need feedback from everyone, with comments and questions. To the best of my ability, I will try to explain what is known and what is not known about the effects of drugs on the body, and especially the mind. I will make no attempt to either encourage or discourage drug use. I will simply try to present factual information. Warts and all.

THE NATURE OF DRUGS

DRUG:

The definition of a drug is probably one of the most difficult of all. Some definitions will be extremely narrow, requiring that the drug be a chemical that is used in man for the treatmnent of some illness. Others will widen the scope to include anything and everything that in some way affects any living system. One often hears the ultimate cop-out: "YOU know what's meant by a drug!" Yet others will try to define by presenting examples of what they feel to be drugs. But in all of these approaches there are contradictions. As examples, I feel that most would accept aspirin as a drug, and most would feel that a glass of water was not. But what about the following shades of grey? X-rays? Cod Liver Oil? Chicken soup? Chocolate? An Exciting Book? A placebo? Vitamin C?

The FDA (Food and Drug Administration) is the regulatory body invested with the authority to approve and regulate the commerce in drugs. It presents the following definition:

The term "drug" means:

A. Articles recognized in the official United States Pharmacopoeia, in the official Homoeopathic Pharmacoppeia, or in the official National Formulary, or any supplement to them; and

B. Articles intended for the use in the diagnosis, cure, mitigation, treatment, or prevention of disease in man or other animal; and

C. Articles (other than food) intended to affect the structure or any function of the body of man or other animals; and

D. Articles intended for use as a component of any article specified in clauses A, B, or C of this paragraph, but does not include devices or their components, parts or accessories.

Medical devices have been specifically excluded from the drug definition. And foods are excluded from the definition of drugs, by the following rather circular definition:

A. Articles used for food or drink for man or other animals

B. Chewing gum

C. Articles used for components for any such article

Professor Samuel Irwin, at the University of Oregon, has given a much simpler definition of a drug:

1

A drug is any chemical that modifies the function of living tissue, resulting in physiological or behavioral change.

I would make the definition looser yet, and considerably more general. Not just a chemical, but also plants, minerals, concepts, energy, just any old stuff. Not just changes in physiology or behavior, but also in attitude, concept, attention, belief, self-image, and even changes in faith and allegiance.

A drug is something that modifies the state of a living thing.

In this guise, everything outside of food, sleep, and sex can classify as a drug. And I even have some reservations about all three of those examples.

DRUG ABUSE:

The problems that were bothersome with the definition of the word "Drug" are nothing compared with the ones that are to be faced with the word "Abuse."

In the context of drugs, the dictionary limits the definition to a single example: Improper use or treatment, and the specific synonyn given for "drug abuse" is misuse. And misuse is in turn defined as incorrect or improper use. The implication is that there is a right way, and a wrong way, to use a drug. This says nothing about the drug itself, nothing about the reason for its use, but simply speaks to the way the drug is used. I have collected over some time a number of examples of the use of the term "drug abuse", or now, more broadly, "substance abuse" as found in articles, essays, or public talks. There are four entirely separate classes for the definition of the term "abuse", and they can be organized around the four operative words: WHAT, WHO, WHERE and HOW.

WHAT DRUGS ARE USED?

This is the only category that assigns the term "abuse" directly to the drug itself. The position taken by the Food and Drug Administration is simple. If the drug in question has been approved or has been exempted by them, then its use should follow approved protocols and its involvement in the patient community is clearly "drug use." If the drug has not been sanctioned by them, then any use is "drug abuse." Thus, there are OK drugs, and there are abuse drugs. Just look it up in a book, and find out which one yours happens to be. I am reminded of the opening slide presented by Dr. Gene Schoenfeld at one of his lectures on drug abuse. It showed a capsule sitting on a flat surface, with a mallet about to smash it to bits. "That," he said, "is true drug abuse."

This concept of good and of bad drugs has many cousins in pharmacology. Some people classify drugs as toxic or non-toxic,

2

or as dangerous or non-dangerous. In fact there was for several years a government agency (now the Drug Enforcement Adminis-tration) that was actually called the Bureau of Narcotics and Dangerous Drugs. This assignment of an intrinsic abusability, or toxicity, or dangerousness, to some drugs and not to others is nonsense. Under certain conditions, all drugs are abusable, all are toxic, and all are dangerous. And all, under certain conditions, can be used sensibly and safely, without toxic effects and in a way that is free of danger.

The FDA classification may be simple, but it is useless.

WHO DICTATES THEIR USE?

A second common definition of drug abuse is the use of drugs outside of a physician's purview. If an MD says "Use a drug for this or that purpose," then that is drug use, and if one self-medicates, then that is drug abuse. A quotation from Dr. Thomas Szasz, in Ceremonial Chemistry:

"Society's prevailing view -- is that being medicated by a doctor is drug use, while self medication is drug abuse. This justification rests on the principle of professionalism, not on pharmacology. [This] concept of drug abuse symbolizes scientific medicine's fundamental policy that laymen should place their care under the supervision of a physician. This is similar to the belief, prior to the Reformation, that laymen should not communicate directly with God but should place their spiritual care under the supervision of a duly accredited priest. The self-interest of the church and of medicine in such policies are obvious. These policies also relieve individuals of the burden of responsibility for themselves."

WHERE ARE THEY OBTAINED?

The illegality, or illicitness, of the source as well as the para-medical aspect of drug use is "abuse" in some authorities' eyes. Dr. Jerome Levine, at the NIMH, has defined drug abuse as:

"The self-administration of [drugs] by individuals who have procured or obtained them through illicit channels, and/or in medically unsupervised or socially unsanctioned settings."

The clear feeling here is that if you interact with drugs in a manner not sanctioned by law, or against the accepted social philosophy, you are abusing the drugs. This, in the final analysis, makes the legal structure, or any generally accepted dictator of social mores, the source of the definition of drug abuse. And since the laws change continuously, and the social structure can vary so completely from one area to another, or from one culture to another, this definition also is useless in addressing the problem.

3

HOW ARE THEY USED?

I personally believe, most strongly, that in the improper use of drugs lies their abuse. Dr. Irwin has phrased it thusly:

" [--Drug abuse is] the taking of drugs under circumstances, and at dosages that significantly increase their hazard potential, whether or not used therapeutically, legally, or as prescribed by a physician."

The medical community largely agrees. The Advisory Panel on Alcoholism and Drug Abuse of the A.M.A. Council on Scientific Affairs defined the following terms:

"Use -- the taking of drugs pursuant to proper medical indications; Misuse -- the taking of drugs for non-medical indications; and Abuse -- drug taking that infeferes with a person's health, or economic or social functioning."

But even this is a little self-serving, in that the perjorative term "misuse" is still the blandest category that functions outside the control of a physician.

People use drugs, have always used drugs, and will forever use drugs, whether there are physicians or not. I prefer the ideas of Andy Weil who holds that drug abuse has nothing to do either with the legal or medical approval or disapproval of the drug involved, or with the reasons for its use. Any use of a drug that impairs physical or mental health, that interferes with one's social functioning on productivity, is drug abuse. And, the correlary is also true. The use of a drug that does not impair physical or mental health, or interfere with social functioning or productivity, is not drug abuse. And the question of its illegality is completely beside the point.

In this regard, drug abuse is a person's use of a drug with which he has a poor relationship. This is Dr. Weil's thesis, and I completely agree with him. In my own case, I have a very bad relationship with Nicotine. I smoked regularly and quite heavily for some 20 years, and then in a most difficult process of self denial, gave up the habit "cold turkey." After not smoking for some 15 years, I had the memorable opportunity of bidding a fond farewell to a dear friend on our last night in Paris, she to return to Germany and I to the United States. As we sat over a last Calvados, she suggested we share a Galois Bleu. I declined, she insisted, I accepted, and in two days I was back to two packs a day. Again, after another few years, the difficult process of self denial, and now I do not smoke. I dare not, not even one. I have a terrible relationship with tobacco, and for me the smoking of a cigarette is a personal example of drug abuse.

This addictive potential, expressed here as a poor relationship with a drug, is in all of us and it needn't be

4

restricted to drugs. I was listening to Hal Lindsey's
fundamentalist Christian radio program a few weeks ago, and got
caught up in the program that followed it. On this "revival"
session, a young female spokesman for the Church had a
transformed drug abuser at hand, and he was unendingly vocal as
to the virtues of finding Christ. I can only paraphrase his
testimonial:

> I used Cocaine, I destroyed my life with cocaine, I
> lost my job and by self-image with cocaine. But once at a
> moment of intense commitment, I said, "Jesus, I accept you,"
> and from then on I had no desire, no urge, and my wife,
> having seen the transformation, joined me in Jesus, and we
> are the most one-ness pair that you could ever see. I will
> go anywhere, and talk to anyone, as to the virtue of Jesus
> over Cocaine.

A commitment to an addiction, and with sufficient
reinforcments towards that commitment, constitutes a conversion
that is real. To exchange a total commitment to drugs for one
to Christ, is not a change in style, but simply a change in
dependencies.

HABITUATION
ADDICTION
DEPENDENCE

These three terms have been used very frequently to describe
the pattern shown by people who are chronic drug users.

Habituation: This is a reasonably benign term, referring
to repeated drug use without serious harmful effects to either
the individual or the society. In a sense it can be compared
with the "sweet-tooth" of the person who craves candy.

Addiction: The term "addiction" has fallen out of usage in
the current medical literature, but it is still popular in the
lay press to condemn drug users and especially certain drugs. It
is a psychologically loaded term, implying compulsive abuse
associated with physical dependence. There are negative effects
implied both as to the individual and to the society.

Dependence: This is the term that has now come into general
use to cover both the psychological and the physical aspects of
repeated drug use. A careful and useful essay by Dr. Maurice
Seevers, in the Chemical and Biological Aspects of Drug
Dependence, has been of great use in framing this concept. A
number of alternate meanings of dependence must be considered
separate from drug dependence.

There are many habits, routine behavior patterns, that need
not be associated with the use of drugs, but which in their own
way nonetheless, rob us of some aspects of freedom of choice.
These are actions that verge on the involuntary, but upon which
we have come to depend for our "normal" daily life. And there

5

are drugs which can be used for replacement therapy and upon which our continued health may well depend. This type of dependency (on things such as Vitamins, Insulin, chronic medication) must be excluded from considerations here. And there are uses of the concept in areas such as microbiology, where one can see an adaptation of bacteria to drugs such as inhibitors or toxins, but this is a matter of genetics rather than of psychological compulsion.

A fair working definition of dependence is the establishment of a conditioned pattern of drug-seeking behavior through the repeated use of a drug. The term is usually used in connection with psychoactive drugs, and has been broken into two subdivisions, psychological dependence and physical dependence.

Psychocological dependence is the branch wherein the drug user finds that his psychic response to the drug is sufficiently attractive to motivate a repeat of the experience. In physical dependence, on the other hand, there occur neural or metabolic changes in the body that result in an improved accommodation to the drug, as seen by either (or both) tolerance or withdrawal reaction.

TOLERANCE

Tolerance is a decreased response within a person to a given dosage of a drug, following and earlier exposure to that drug. This declining effect often prompts the increase in the self-administered dosage. It can develop rapidly and has been known, with some drugs and with certain people, to be observable after only a single exposure to a drug. In this case it can be called "Acute Tolerance" or Tachyphylaxis. And when the tolerance has developed over a period of time, following many exposures, it is called "Chronic Tolerance." With some drugs, the increasing of the dose administered (needed to produce the expected effect) will reveal some new pharmacological properties that were not encountered at normal doses. The production of a psychotic state following exposure to high dosages of Amphetamine by the tolerant individual is a well-studied example.

An aside is useful here to bring in the definitions of some of these additional words, since this is after all a section devoted to definitions. "Acute" has a popular meaning implying severe, or extreme. But in medicine the usual usage is to emphasize one-time, or a single occurance rather than recurring; in medical practice, the word "acute" describes a rapid onset of symptoms of a condition. And the companion term "chronic" means that something has been repeated, or experienced many times, again and again. There is an intermediate term meaning not just once, but then, not many times either -- just a few times. This term is "subacute."

Another word above was the term "tachyphylaxis." Tachy is a prefix meaning fast, and phylaxis derives from the Greek word for shield or guard. This suffix is found in other terms such as

6

prophylaxis (a guard against, as in protection against disease) and anaphylaxis (a guard upwards or backwards, representing the medical concept of sensitization).

Metabolic tolerance (or dispositional tolerance) is a loss of responsiveness due to actual changes in the body's capacity to metabolize or bio-transform the drug. This usually follows the stimulation of enzyme systems in the liver that are responsible for inactivating the drug in question. Pharmacodynamic tolerance is a term implying the adaptation of the body (the receptor sites or actual areas of drug action) in a way that decreases the intensity of the response.

Cross-tolerance is the recognition by the body of a new drug following the development of tolerance to a different drug. The two drugs are usually related to each other pharmacologically, and quite often share some structural features.

WITHDRAWAL:

When a person abruptly stops using a drug, there is developed a state of "withdrawal", which is also widely referred to as "Abstinence Syndrome." (The term "syndrome" refers to a group of symptoms that has been used to define a medical abnormality).

Withdrawal from psychological dependence leads to displays of emotion, to exquisite rationalizations (for continued use of the drug that has been withdrawn), and often to some form of alternative pleasure seeking. However, in examples where drug usage has led to physical dependence (physical addiction), withdrawal symptoms can be modest or they can be life-threatening, largely dependent on the drug and on the history of its use (time and amount).

In the instance of morphine and its related agonists (an agonist is a drug that produces similar effects, whereas an antagonist is a drug that blocks the expected effects) the most common physical signs are abdominal pain, irritability, cold sweats, diarrhea, nausea and vomiting. The abstinence syndrome associated with heroin withdrawal has been compared with a very bad cold. This syndrome will occur in a day or so following drug cessation, or virtually immediately following the administration of an opiate antagonist.

With the barbiturates, the syndrome can be dangerous. One sees effects that resemble those that are characteristic of a convulsant drug. There is seen within the first day neural hyperexcitability and tremors. In the second to the eighth day, there can be Grand Mal convulsions with both tonic and clonic character, and mental blackouts. There is commonly a delirium and a psychosis-like state with thought disorder, hallucinations, and paranoid delusions. The generalized weakness can last a couple of months and there is usually a REM rebound.

7

With alcohol this withdrawal syndrome can be more modest (following the acute exposure, the hang-over, or simply sleep disturbance and irritability) or it can be expressed as a full-blown delirium tremens (everything experienced with the narcotic withdrawal, including convulsions, but more often of the Petit-Mal type).

Vocabulary time again. The term "hyperexcitability" means more excitable than usual, and introduces a neat set of prefixes. When something is normal, the appropriate prefix is "eu-". Thus one sees terms such as euthyroid, meaning that the thyroid is acting as it should. (An interesting viewing of our present acceptance of the life around us to be found in the term "euphoria" which by its origin means "normal feeling"). The opposite prefix is "dys-" which means not normal. Thus an abnormal thyroid is called dysthyroid, but it can be too high in activity (hyperthyroid, from hyper- meaning above) or too low in activity (hypothyroid, from hypo, meaning below). The terms "tonic" and "clonic" are often intimidating when seen in reference to convulsions. A simple mnemonic trick can keep them straight. With "tonic" remember tonus which is a term referring the the firmness (the tone) of a muscle. Thus a tonic convulsion is one in which the muscle becomes rigid and distended, completely inflexible. And with "clonic" remember a cyclone in which the weather bashes everything back and forth. A clonic convulsion displays an alternation of contraction and relaxation of the muscle, providing a sort of thrashing action. REM stands for Rapid Eye Movement, a phenomenon involving the coordinated searching with the eyes right and left, up and down, that occurs during sleep and which is felt to be associated with dreaming. With certain drugs such as the Barbiturates, there is a deficit of REM (and thus presumably of dream time) and it is now believed that a certain amount of dreaming is necessary for the mind and body to be in good health. Then, with the withdrawal of a person from dependence on Barbiturate drugs, there is a period of catch-up, called REM rebound, and there appears to be intense and not necessarily pleasant dreaming. And the term "mnemonic" refers to a device (or a trick or a strategy) for helping one to memorize a word, or a series of names. The three Baltic States, in North to South order, are alphabetical, vis., Estonia, Latvia, and Lithuania. From the same source is the word "amnesia" meaning without memory.

CARCINOGENICITY:
MUTAGENICITY:
TERATOGENICITY:

Again, here are three terms which are often confused.

Carcinogenicity mean, literally, the ability to cause cancer. The suffix -genicity is found in all three of these terms, and is often seen as -genesis. It refers to the origin, or "Genesis" of a thing. There has been great concern about the exposure of living things to agents that can produce cancer. However, in fact there are very few things that, when brought in

8

contact with a human or even a test animal, have actually been shown to have this capability. In experimental animals, to generate cancer by repeated exposure to even the most potent carcinogens requires repeated exposures.

The carcinogen must sometimes be balanced on a fine line separating drugs from not-drugs. An example is the classification of the compound Saccharin. This is a non-nutritious sweetening agent, which is known in certain instances to produce bladder cancer in mice. There is an amendment to a food law that prohibits the inclusion of any drug as a food additive if if produces cancer in any animal in any amount. This problem was resolved by renaming Saccharin as a drug rather than as a food additive, hence it may be included in foods, as there is no law that prohibits this.

Mutagenicity, on the other hand, describes the creation of a mutation, and is only suggestive of causing cancer. The Ames test for mutagenicity depends on the induction of some metabolic change in a microorganism by the chemical in question. A strain of bacteria that cannot grow without X but gets it from its food supply. If a chemical or extract that is free of X is given to quiescent culture, it will not grow. But if the food causes the organism to mutate (with some insult to the genetic structure of the bacteria) and to become able to produce its own X, then growth will occur. And the extent of growth is a measure of the potency of the chemical as a mutagen. There is no certain connection of this action to carcinogenicity, but the fact is that some chemicals that are mutagens do indeed cause cancer. Many do not.

Teratogenicity is a completely different matter. Terato- is a prefix that means monster, and the word stands for the malformation of a fetus as it is developing in the uterus. There is a period of sensitivity during pregnancy, in humans somewhere within the first several weeks, when the developing child is exquisitely sensitive to exposure to foreign chemicals. A tragic example is the development of a phocomelus (a name derived from a seal's arm) by prenatal exposure to the sedative-hypnotic Thalidimide. A general rule should be inflexibly observed: use no strange chemicals (foods or drugs) during pregnancy, especially during the first trimester (first three months).

LETHAL:
TOXIC:
POISONOUS:

Three more occasionally interchanged word. Lethal is the most straightforward. This is the level of a drug that will kill a living thing. A common, and I feel largely useless, pharmacological procedure is the determination of the lethal level of an experimental drug. This is the measurement of the LD-50 of a material, a term that means the dosage level that will kill half of the animals to which the compound was given. There

9

can be an LD-95, or an LD-5, representing the lethality of a
given dosage to 95%, or to 5%, of the test animals. The term is
used to predict the potential lethality of the drug to man,
perhaps only by comparing a series of compounds to determine
which may be the least threatening.

A related measure is the effective dose, the ED-50, where
half of the test animals will show some desired effect. When one
looks for, say, a sleeping aid, one can measure the recovery of
the righting reflex in a test animal to determine the
sedative/hypnotic potential of a drug, but when the effects
sought are of a psychopharmacological nature, animal tests of
effectiveness are indirect at best. The quotient of the two
values is called the Therapeutic Index (T.I.), and it is obvious
that a drug manufacturer would like the value to be as high as
possible, if for no other reason than to provide a safety margin
in cases of overdosage. A value of 1000 is wanted (the lethal
level 1000x the effective level), 100 may be tolerated, and yet
some drugs in common use (such as alcohol) have a T.I. of only
about 5.

The term toxic refers to any undesirable effects that may be
produced instead of a desired response or along with the desired
response. A material that produces only undesired effects is
called a poison.

HALLUCINATION:
ILLUSION:
DELUSION:
FANTASY:
IMAGERY:

Here is a collection of words that are easily used
interchangeably, and which mean many things to many people.
There is no standard definition, so the best I can do I to
describe how I use the terms.

Hallucination is simply the seeing of something that others
believe not to be there. The term "hallucinogen" is easily used,
both in legal language and in the popular press, to describe any
and all compounds that provoke (or allow) changes in the visual
field. And many of these so-called hallucinogens do indeed lead
to perceptual alteration, but very few of them inspire the seeing
of things that are created de novo (with no point of origin).
Most visual syntheses are distortions rather than creations.
True drug-induced hallucinations are rare.

There can be a subclassification of these visual distortions
into retinal as opposed to associative distortions. A retinal
distortion involves the fine detail of the macula lutea, the
small yellowish spot to the side of the center of the retina
where there is the maximum visual acuity. Look at a fine
straight line. There is no exact straight-line arrangement of
cells in this high-resolution area of the eye, so what occurs is
that the image of the line activates the cells nearest to the

10

image, some to one side and some to the other. The other eye does its thing similarly. And it is the associative area of the visual process (near the back of the head) where there is the choosing of which cells of which eye to respond to and what data to use, since somewhere you KNOW that the target to be seen should be a straight line. But with some drugs (which cause distortion in this detail) the choices are different, as the intended target is not clear, or comes from some unconscious source. Expand that straight-line concept to a complex image (as with the surface of the running wax from a candle) and strange shapes can be seen. Such an image rarely can be maintained for more that a fleeting moment.

The associative visual distortion is closer to the process better called an illusion. There is an interpretation that comes from the visual process, an assignment of identity or meaning to something seen, that is inspired by some aspect of light or color or motion, and which is confirmed by some reinforcement from your own imagination and response to it. Sinister faces may be seen in the ivy leaves, or a particular flower may glow and call you to it.

Auditory distortions are more commonly seen in spontaneous mental illness, but are correspondingly rare in drug experiences. The hearing of voices (the Devil told me to do it) is the province of the psychotic. More often associated with drugs will be amplification of sound, the selective hearing of one component of a mixture of inputs, and occasionally a distortion of pitch or of timbre of a voice or a musical note. Modifications in the other sensory inputs (such as the sense of smell) are extremely rare.

A delusion is the interpretive state that can follow from a combination of the above distortions. It is an impression, a point of view, that instills itself as a belief which has no focus that can be confirmed. No one else can truthfully document it, but yet it can haunt you as a feeling that you unshakeably know to be true.

Most of the above distortions are associated with the visual process. With the eyes closed, most of these phenomena cannot exist. The last two terms above, fantasy and imagery, are part of the human experience that is separate from the visual process.

Fantasy is the "Where you are" of the imagination. One can construct a dream-like world in which you participate, with associative interactions and imagined voices and imagined contacts. All senses, except for the visual process, can be involved in this created reality. Imagery is the "What you see" companion in this eyes-closed world. Here one can construct shapes, castles, arabesques, and assorted visualizations often with their origins in sound inputs such as music. Both worlds can occur simultaneously, with an effective closing off of the reality about you.

11

Combinations of these mental phenomena can be part of a drug experience, and they can occur spontaneously. No one is a stranger to these worlds, as they occur to every person daily in the state known as sleep.

This is an introductory outline of some of the vocabulary that will be used and added to, during the course of these lectures. I will need feedback from everyone, with comments and questions. To the best of my ability I will try to explain what is known and what is not known about the effects of drugs on the body, and especially the mind. I will make no attempt either to encourage or to discourage drug use. Just the facts, M'am, warts and all.

12

February 5, 1987

The Origin of Drugs

United States Drug Laws and Drug Enforcement Concepts and Agencies

[The first sentence of Sasha's opening comment is missing due to ambient noise.]

SASHA: It is not an effort to get all of the vocabulary in one place. Good heavens, you can get a dictionary, or medical dictionary, for that. There are some of these words that I'm going to be using again and again through the course: dependence, tolerance, acute, and chronic. And as we go on, we'll be getting a lot more terms as we get into various disciplines in more detail. More words will come up. But these are some of the basic ones that go throughout the whole drug discussion. Are there any questions in general about what is going on?

STUDENT: Oh yeah. Methadone. Is that as physiologically addictive as heroin?

SASHA: Methadone is very comparable to heroin in its physiological and psychological addiction, yeah. Methadone is a totally synthetic drug. Interestingly, I don't know if I'll get into that. I probably will. Methadone was originally known as Dolophine. It was created in Germany. A lot of the incentive during World War II, in the development of chemistry and some of the development of drugs, was an outgrowth of the fact that this philosophy was fighting that philosophy. But this philosophy had access to the raw rubber; that philosophy had access to the opium. Each side had needed rubber, each side had needed opium, so to speak. So, there was a great drive here to find some way of bypassing the need for rubber that came from the southern part of Asia. And in Europe there was no immediately easy access to opium and there was a great drive to find drugs that would have the morphine-like action. Dolophine was developed as a morphine substitute and was named in respect for Adolph Hitler. Dolophine from Adolph. An interesting little bit of trivia.

[Directed to student.] Yes?

STUDENT: Why is it preferable to have a junkie hooked on methadone rather than heroin?

SASHA: I don't think it is. The main rationalization that's given is that methadone is legal, and we can prescribe it and we can control it and people in medicine can administer it. Heroin is an ugly, illegal drug and you're not allowed to touch it under any circumstances. So you're replacing a "cannot handle" with a "handle," which pharmacologically are very, very similar.

[Directed to student] Yes?

STUDENT: But why is methadone legalized over heroin?

SASHA: They never illegalized methadone. In fact, we're going to get into that later on in the course of this lecture today, in the history of the drug laws. The law sets up these structures and once they are in there, they remain, even though they may become archaic.

For example, I was born in Berkeley, I was raised in Berkeley, I went to school in Berkeley. When I was small, they had little streetcars that bounced around going down tracks—there are no streetcars in Berkeley anymore—and the front and the back porches were open. Not porches, but where you get on and you're not inside yet. That outer area was open. And there was an ordinance that had been there since the very founding of Berkeley that it was illegal to shoot a jackrabbit from a streetcar, an ordinance that's still on the Berkeley books. There are no longer any streetcars and I don't think there are any jackrabbits, except maybe in the hills, but it's still illegal.

Here's another example. It is illegal to go through San Francisco on your way to Sacramento to see a federal marshal. Now there's a reason for that. If you have a person coming through and you don't like their looks or you don't like the way they're walking or you don't like the appearance of their car or something, if you suspect that they're on their way to Sacramento to see a federal marshal, you can hold them for seventy-two hours. And if it turns out they're not on their way to Sacramento to see a federal marshal, let them go.

It's known as a holding law. It's a device, it's a stratagem by which you can take a person and check on this, check on that, what's their background, are their fingerprints wanted, is there a flier out about them. You can't just go out and say, "I don't like your looks. I'm gonna bring you in and hold you for seventy-two hours while I see if you're wanted in Tuscaloosa." But you can say, "I suspect you're on your way to Sacramento to see a federal marshal. Seventy-two hours." Some of these old archaic laws have a reason, but the reason can't be contained explicitly within the law, so the law is made in sort of a circumventory route.

When we get into talking about the laws on drugs, you'll find an extraordinary litany of laws that have become more restrictive, more intense, more supervisory, one after the other after the other, because the goal is to stop drug abuse, to stop the illegal traffic in drugs. And that goal is going to be achieved by increasing the fire power of the law. Somewhere along the line it is going to be apparent that the goal is not being achieved, but we will have a pile of very, very restrictive laws. We already have a pile of very restrictive laws on the books, which I want to talk about in the latter part of this hour, and which cannot easily be undone. Laws are not easily undone.

This is again an aside, but if a person in power makes a statement—this is kind of a general term for people in power, be they lawyers, even more so in politics, be they in administrative office, small things, university, big things, the government—if they are in power, the reason they're in power is they have acceded there over someone else and they're more powerful or they're more manipulative, they're stronger, they've got more money, they've got more books, they have more friends, they have more clout. That's how they've gotten there. And one of the reasons you stay in power if you're in power—you can see this conspicuously in some of the minor dictatorships around the world, but you can see it a little bit more covertly in some of the major countries in the world—the way you stay in power is not to make mistakes. You'll do something; it is not a mistake. If it's a mistake, then you did something wrong and someone who's not going to make a mistake if they get into power will sort of nudge you out. And so, if you are in Congress, you pass a law; if you are in the Executive Office, you do something. Whatever it is, it is not a mistake. And if necessary, you will change reality or you will deny something, or you'll modify something to keep it from being a mistake.

As a specific example of this, very recently a law was made that put a compound commonly called "ecstasy" into the Schedule I drugs because it had no medical utility, it was purportedly unknown to the medical community, and it had wide abuse potential. Well, the truth is it was known to the medical community, it had wide medical use, and there was no conspicuous evidence of abuse potential. But the statement was made. The people who made the statement did not know these things, they are not wrong, and hence the law placed it as a high abuse potential, no medical use drug, which is the way it will stay. Because there is no error made. Once you make an error, you're displaced by people who are waiting in the wings to catch you stumbling and, as with the survival of the fittest, will elbow you out and take your role.

So, you'll find a lot of these little laws are inflexible because they were made for good reasons and they can't easily be undone. To undo a law, to rescind a law, to withdraw a regulation is, at some level, to acknowledge that that law or that regulation was not appropriate. And look in the area away from law, look at the area of religion. Look at the stand that has been taken, say, in the church, in any of the churches, where this is dogma, this is totem, this is taboo, and that will be maintained for maybe a few centuries and a change of personnel. I think recently they are considering apologizing to Galileo,[1] but it has taken a while to let that kind of thing evolve.

I want to get into, initially, the history of drugs, and then into the history of the law. The origin of drugs is—well, we're back to the origin of humanity. I've always had trouble myself trying to visualize what it would be like if I lived way back 5,000 years ago when history was just being laid down and people were respecting kings and building pyramids and possibly taking camels across the desert. I try to visualize myself in that role because if I really understood where I was in that role, I might be able to understand that aspect of history. And I've always had the fault of trying to place myself in that historic role with the knowledge and with the sophistication and with the facts that I have at my disposal now. As if that were a transformation that could go back and forth. And I never have been quite comfortable with the reality that that whole transition from then to now has been one of irreversible change. When something occurs, something has changed the conscience or has changed the appreciation of reality to the human. That change is there forever. You can't undo it and go back and try it another way. The entire progress of the development of the human has been one of irreversible change. And we see that yet today. When something occurs today, you say, "But my heavens, could we try it some other way?" You say it in some mental, unconscious way. Could we try it another way? We can't. Every single thing, every single act, is in the books forever.

Even more than that, don't go back 5,000 years to the beginning of recorded history. Go back to the origin of humanity because that's where drugs really started. And that's back, what, a hundred thousand years, a million years? Now I'm really going to get rhapsodic and hand waving because I don't know. No one knows. There is no record. All you can do is try to guess what happened at the very origins back there on the basis of the hints you can see with groups perhaps still intact in the world that may not have changed much since that time. Study the aborigines, study

---

1 Pope John Paul II finally did so on 31 October, 1992. Galileo had been condemned in 1633.

the people who have been cut off and have not been enlightened and have not received the benefit of our own transitions. Try to glean from them what happened. Put yourself back into, let's say, a million years ago. Something occurred. Human beings as we know human beings, let's say the organism that we know, maybe with a lot more hair, perhaps shorter and certainly more simple and barely able to talk; but human beings as such. Somewhere there's a transition from ape or animal or not human, whatever that transition is. It may have been by divine—we're going to get into that aspect in a moment—generation. It may have been by an evolutionary process, but somewhere in the process there occurred, maybe in one place, maybe it oozed over a period of generations, maybe it occurred in some divine moment, maybe it occurred in one place and it spread through the world, maybe it occurred in a thousand places over a hundred thousand years, but somewhere an animal, such as the human, evolved and had the capacity to ask, "Who am I?" That animal had the capacity to realize that others have died, and I will die, and was able to communicate other than by very rudimentary attention-getting language concepts.

Put yourself at your own age in a small group. There is no fire, there are no clothes, there is a very hostile world around you, eat or be eaten, find food somehow. If you undergo this or this or this damage, it's known that you will die and you're afraid of death. That fear of death is part of this. You ask, "Will I die?" "Why will I die?" "Where will I go when I die?" This kind of intellectual curiosity is unique to human beings as far as we know. Perhaps guinea pigs have it. But as far as I know, it's unique to human beings.

All these questions. They're the same questions we're still asking today. "Where do I come from?" "Where am I going to go?" "What happens when I die?" Then you get down to these rudimentary drives: to stay warm, try to get warm, to stay fed. How do you get food? You go out and you try eating this and you go out and you try grabbing that. How do you get an animal? Well, maybe all of a sudden—I think you've seen the scene in *2001* where they touch the big stone and suddenly pick up a club and club one another—it might have been that type of transformation. But the mechanical coordination was there. And the intellectual curiosity was there.

So your age—you are pushing twenty-five. You are in the land of the aged and not many of your peers are still alive. They have mostly gone their way to wherever that is. You ask, "What is it in me that goes when I die?" You get this idea that someone, somewhere must know, and you ask. Whom do you ask? You ask the person who has this knowledge and is conspicuous because he wears things that make him different. Everyone has the same number of arms and ears and appendages and such. But there is someone who is the shaman, the person who's the knowledgeable seer. He or she is the one who has somehow perpetuated the myth of where things have been. Read the stories of the "Dreamtime" of the aboriginal Australian. This is a superb example of the idea of explaining the origins through myth and through artifact and through story, all requiring the verbal language. There is no written language and if you were able to look for it, you'd see there would be no written language for maybe half a million years. Maybe you can scratch little things. But what do you scratch? You scratch something perhaps to embody the spirit of what you're handling. You have a soul, and the soul term is our recent term. But you have something in you. And so does the tree and so does the child and so does your mate and so does the animal you eat and so does the bush you take the leaf off of. All these are souls.

Sickness is there. However, sickness is one of the few things you can't point to as being a soul. You can see a tree and you can see what happens to a tree when it falls. But you can't see what happens to a person when that person begins acting strangely, or the person has a toothache, or the person has a broken leg. Where is that aspect coming in? There was the development of gods—gods came later—the development of images, development of demons. If you had a toothache, you had a demon that was a great big tooth that was probably colored red. And if you had an earache, the demon had big ears. You couldn't see it, but you knew it was there because there had to be some explanation for illness, and for disaster, and for good things and bad things. They all had their demons. And there was a shaman, a medicine man or woman, who later became a priest(ess) or a physician, but they're really the same role, who had a way of communicating with that demon, finding the demon. The person who's lost their soul, where is the soul? Well, the shaman knows it's in a hollow that is known to be in a tree that's over the hill and down there. And they will bring that soul back and pound it into the person.

Look how one handles what we call mental illness, a very nebulous, shifting definition. My mental illness is not your mental illness. Go to another culture. You get a different definition of mental illness. And to another generation, a different definition of mental illness. When a person is not sensible and you know this person in some way is a threat, how do you repair that lack of sensibility? Well, the shaman will come over or you'll go to him. They know the answer. They will communicate. I don't know how they understand what they are talking to, but in some altered state.

The concept of altered state, by the way, you're all very familiar with, at least at that age. Because even at this age you spend six or eight hours a night in an altered state. Your whole sleep pattern is a dream world, and your present world cannot really recall that dream world. Some people are very strange, and they can recall dreams with fine detail. I sometimes suspect they're on something. [Laughter.] But, I know in my case I can't recall that dream world. But put yourself in that dream world, and in that dream world you can't recall your real world. You are in a world that has no touch of what you think is your real world. Which is the *real* real world? There is no absolute.

And this billion-million-year-old group, this tribe of which you are a member, has this dream world. This shaman can use that world. They may eat a leaf of a plant. Remember that the treatment of illness with medicines—and we're talking about drugs—at that time, it was not the person who was ill who used the drug. It was the shaman who used the drug, and the drug gave them insight as to how to cure, or how to find, or how to remedy, or how to treat, or how to appeal to the person who was ill, or had a broken leg, or a toothache, or was insane, or dying. The shaman, the priest, the knowledgeable leader, probably the only one who had an IQ that was pushing sixty-two. What was the intelligence at that time? We have no way of knowing.

But it could have been, and probably was, pretty doggone elementary. What was the sense of self, the sense of ego, that marvelous little transition that occurs, I don't know if you remember, somewhere around eight or nine or ten years old, that prepubescent, preadolescent time? If you can recall back, somewhere usually most people at around eight or so go through a fantastic transformation in which they go from the sense of "me" to the sense of "I." Where suddenly it's not "He hit me," but "I was hurt." And that person evolves, that ego sets in and it's the prelude to the whole adolescent thing that has its own evolution in its own way. But that little transformation occurs to us perhaps at the age of eight or ten. I think mine was around nine or ten. I never quite saw it until it had happened, and I said, "My golly, where was I for eight years?" I think you all kind of have a feel for that. That may not have occurred for early people, or it may have occurred at a later time if at all. So, I think with your earliest leaders you had this strong person, the person who had the personality to go out and find the animal for dinner. This time was all hunting and gathering.

If you had a broken bone, it was wrapped in a leaf. What would you do with a wound? Here was a wound with blood gushing out. Dismiss the knowledge of medicine, dismiss the knowledge of drugs, only know you can appeal to someone who's wearing a wolf skin, a chain of wolf teeth around their neck, and has things hanging from their belt and a drum and a little pouch that's full of miracles, full of magic. You go to this person with a pouch full of magic and say, "My brother is bleeding badly." And they'll come over. What would you do to a bad, bleeding thing using the rudimentary intelligence you have? Wrap it in something. Stick your hand on it. Put your finger in the dyke. You know this bucket of blood, you know it's blood, you've killed animals, got blood everywhere. You're full of blood. No knowledge of circulation. That came a hundred thousand years later. You're a bucket full of blood. As you put a hole in a bucket full of blood, blood comes out. You put your finger in the hole. Well, you can't go around all day with your finger stuck up someone's hole. What are you going to do? Wrap it in something. Well, a leaf, but what leaf? That's a soul, too. That's something living. And the shaman says, "I will find it." And they bring the leaf. And after a thousand tries, here's a leaf that minimizes infection. For some reason that leaf did something or it made it numb or the crying stopped sooner. This leaf made the crying go longer and caused the loss of an arm and the loss of a person. And over the millennia you begin getting that relationship with nature that says, "Here is something that helps. Here is something that doesn't." These demons, the toothache demon, the earache demon, the person's soul demon.

After the hunting and the gathering became a time of beginning to bring the crops to where your tribe lived rather than taking your tribe to where the crops grew, the beginning of agriculture, the beginning of the awareness of the stars as something that bore a relationship to when you planted and when you harvested, and when you gathered in and when you let out. The seasons were instrumentally locked to this—you're living on a sandwich. Here's the earth and there's the sky. And the sky is moving past the earth in some miraculous way. But the sun moves low or the sun moves high. When the sun is moving higher, you plant. When the sun is moving lower, you harvest. And then you save until the sun does this again. Who follows this? This is memory. This again is the knowledge of the magician, of the person who follows such things and has that intelligence to make the connections between the passage of the sun high/low, a long day, a short day. Even the dictations of how you bring the seasons back, the festivals where you kill someone who is dear to you to guarantee that. These ideas of ceremonial givings are still with us today.

I think one of the most beautiful examples, one of my first big, real lessons on how people must have, in primitive times, actually killed someone—consider the concept of killing one of your tribe. Now that is not seen in an animalistic world. You'll find exceptions from the black widow to occasionally a wolf pack. But in general, the killing of a member of the tribe is one of the taboos. It's built in there, and yet, there are instances where killing must be done because there has been a taboo that has been broken that is so devastating, so threatening to the tribe, it's better to lose one than to jeopardize all.

Here's a specific case, it's a case of a drug. They classify this as a drug in the Goodman and Gilman Pharmacopeia. It's a piece of bamboo and it's about three inches long. I don't know botanically what you call a little node in the middle of a bamboo where water doesn't run through;[2] but, if you cut above and below the node you have something that's discontinuous. And you can fill the top and you can sharpen the bottom. And you take this node, three inches long, about an inch in diameter, with a node in the center, and point the bottom down. In the top you put a mixture of three things: There's the ash from a given tree, there is a soil from a certain area, and the third thing I've forgotten, but I have it in my notes somewhere at home. And you mix these together and you put it in the top of the bamboo. And then this piece of bamboo is put in the pathway in front of the door where the person who's been condemned to death lives. If it's been a minor condemnation to death, you have degrees of this, you banish them from the tribe. The person has to go out. They cannot have any support from the tribe. They die. Outside of this enclave of people who can barely eke out their existence as a group, an individual does not live. If they find another tribe, they'll probably be executed by them as being not part of them.

But, if it's within the group and you must be assured of the person's death for the preservation of the tribe, you put this little piece of bamboo in front of their door. And as they come out of the door, they see this bamboo and realize they have been capitally condemned. The person has been condemned to death. They'll go back into their cabin, wherever it is they sleep and live, and in twenty-four hours, they are dead. The person dies because they know they cannot live. Now this is a beautiful picture

---

2 Here, Sasha is referring to the bamboo joint itself. This is called the node, the part in between is called the diaphragm. The material comprising it is referred to as the culm.

of what may have occurred a hundred thousand years ago, but the truth is, it occurs today in the Fore people of New Guinea. This is the way they execute. By taking bamboo, putting a mixture in the top, putting it in front of the door; the person sees it, and they die. Very effective form of execution in a tribe that has no capability of hurting. They cannot inflict punishment. They cannot kill their own. They execute in an absolutely nonviolent way. Let the person do him or herself in. The person does not kill him or herself physically. They let their life go. They have been so brought up in that knowledge that this works.

In our current way, we have just as many rudimentary instincts in us. We have also been so brought up in that knowledge: The medicine man or woman must wear a necklace of tiger teeth and pouch alongside. You go into a hospital and try to get advice from a physician who's not wearing a white coat and a stethoscope and a name tag, but some person in jeans wearing an old shirt with the tail hanging out, wandering around in tennis shoes. Somehow that person doesn't have the authority to really help me when I'm sick. You want the image, you want the picture, you want the uniform, you want the medicine man or woman. Go and talk to a minister if you are deeply involved in any particular religion. You talk to a minister who's not wearing the symbols that gives that person the stand-apart-from-the-rest-of-us-ness and you don't have that authority. We still carry all these things right with us.

This knowledge of plants, so-called green medicine, to a large measure, we now know that some of these are effective and we believe some of them not to be effective. But you must remember the placebo effect. Those things that are believed to be effective tend to be effective. Different cultures, different plants. But there was this development over hundreds of thousands of years, usually directed toward the shaman, and usually towards the shaman's ability to get knowledge, to get insight, to get fact in a way that no one else could. That's why that person was the shaman. He or she was able to communicate with the demons.

With the development of agriculture, the demons were removed from the earth and became located in the stars. Hence, we have the concept of the heavenly bodies. The stars and the signs of the zodiac dealt with planting, and also dealt with illness and tragedy and happiness. All these things came from outside, other souls outside of us, to us. You'll find, as you get into the recorded history, records to the orientation of the stars and to the gods, often with animal faces. Look at the history of the development of medicine in Egypt where you had god symbols that were health and happiness and death and life. All these aspects were outside of the person, but were demons or gods. They're interchangeable.

In Greek and Roman times, now you're into recorded history and you can go back and find that they used this as poison and they used that device for curing, and they used that device for hurting. They used that device for diagnosing. They cast the entrails of a chicken and how they fell told where the damage was or where to go. These are structures. Almost no drugs except for those that were used by the priestly cast, and by the few. And, in many ways, that still is part of our structure today.

You tend to say, "Well, gosh. We have knowledge that disease is caused by microbes and you can kill it with an antibiotic." Cast yourself a hundred years ago. Microbes were known 300 years ago. A hundred years ago they didn't know microbes caused diseases. Surgery was still run with nice bloody gloves, high hats, ties, and jackets. The concept of infection came from the air, from the spirits. Take the word "malaria," those who've delved into French—*mal aire*, bad air. You know you go in the swamps and you come down with a horrible disease that was due to the bad air. Malaria.

There are cultures here and there around the world that had their own systems. The South American culture, the Asian culture, but the main record we have followed and consider to be our record, the one that has come up to where we are now, is the record that's involved the Mediterranean area, from Africa to the Near East to Europe. The so-called Cradle of Enlightenment is through the Greek and Roman times, from Egypt to Greece to Rome and into Europe. This is a little bit self-serving because there were many, many absolutely independent and totally separate areas that also existed.

One development during the Dark Ages was the development of the concepts of alchemy. Really the first explorations of what is there. Remember that atomic theory, the reduction to atoms, back then could only be explained in terms of the mathematics of the ancients. At that time alchemy was developed into a very fine art. I've drawn this on the board. It gives the spirit of what was going on. We know, from their point of view, that there were fundamentals out of which everything was made. This is a particularly nice one because it's one of the structures of alchemy where you have dryness and dampness and warmth and cold. And you have the four elements out of which everything was made: fire, air, earth, and water in some proportion. The personal character of any person was built on this and the deficiency or excess of that character was associated with it. If you were depressed: the cold earth. Confidence: air rising, the lightness. Fire and the intensity of anger. You can see the deftness of these associations.

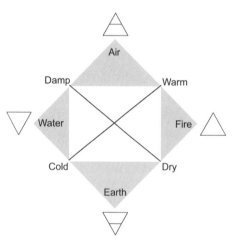

There was a search. What were the alchemists looking for? It's easy to say they were looking for gold and wanted to transmute lead to gold. This is what was said, but this was not really what was going on. The transmutation was wanted, but it was transmutation of themselves. They wanted to find how they could interact with things and in a sense find whatever would allow them to understand themselves. Perhaps to find eternal life, perhaps to find ways of maintaining their own image past its mortal limit. We have the same urges today. We write books because we want something to last beyond us. We have families and raise children because we want something to last beyond us. Because at some point the knowledge that we are mortal does come across and you realize there is only just this much time that you have. And in that time, you must leave something that will keep you alive forever. That aspiration to immortality is one of the human virtues. That was really their search.

These are the four essences. And one of the things they were looking for was, in essence, the fifth essence, the one that was the soul of all of this. We have the term in English today: quintessential. Which is the substance that puts everything into place? The quintessence, the fifth essence. That was what they looked for in the days of alchemy. And to find a form of gold that could be drunk. Gold was the image of light. There were at this time six or eight basic metals that had been discovered by trial and error over the centuries. And it turns out the six basic or eight basic metals correspond with the six or eight basic moving things in the sky. There was the fixed sky, the stars. There was the moving sky, which was the planets and the moon and the sun. And the brightest thing in the sky was the sun. The most precious and the rarest of metals was gold. The next brightest thing was silver and was the moon. And the connection between the metals and astronomy has lived until today. Our word today *aurum* for gold and for the sun. In fact, in my lab at home, I often use old alchemical names for things. Lunar caustic. What is it that's associated with the moon and corrodes, is caustic? Silver nitrate. Silver, second metal, lunar caustic, silver nitrate. It's a sterilant. It's been used for years because it stops disease. You don't know what disease is due to, but invoking the religion of the moon is somehow effective in stopping disease. The sun doesn't do it. And you have iron and warfare [Mars]. And you have lead and Saturn. And all these things tie together in really a remarkable way.

Sometimes I almost get into a strange spot wondering if we're looking not at happenstance of effect, but at some form of causality. Sometimes I wonder if a person interacts with something and gives it character, that character is embedded in what they are interacting with and is seen by other people as that same character. So there may actually be a touch of doing-ness that has brought some of these relationships together.

The alchemists combined what they could. What was available through the time of alchemy, in the 1200s, 1400s, 1600s? The things that could be distilled from the earth, could be distilled from plants, distilled from minerals. Fire, heat, the various structures of the alchemist's laboratory were basically those of heating, cooling, driving with temperature, solvent extractions, the search for the universal solvent, the search for the insoluble. All this laid the framework of what eventually became chemistry. At the time, the first evolution of ether is a very good example. Diethyl ether was one of the first anesthetics that revolutionized the practice of medicine. It was first made in 1200 by the treatment of old wine with sulfuric acid. When it was being distilled, what was left was ether. And it was known at that time to be numbing and to be sweet. Ether was called sweet vitriol, vitriol from vitriolic acid. Vitriolic acid being sulfuric acid which is sour. Ether was the sweet liquid that came from vitriolic acid and wine and was insoluble in water.

Wine was available. There was a stupefactant known as alcohol. How was this first discovered? Probably someone back in the old hoary days a hundred thousand years ago was wandering out three days after a rainstorm, the water had gotten into a beehive where there was a lot of honey and the honey had diluted and started fermenting and they tasted it and fell over kind of giddy. Next thing you know, all their friends were over there tasting this diluted, older honey. [Laughter from class.] I don't know how alcohol first got in the stream, but every culture, with about two exceptions, has had some form of alcohol. Every culture has found plants that have numbed, have put to sleep (your opium), that have excited, have made sloppy (your alcohol), you have your marijuana, you have your mandrake, the stupefaction of mandrake, and some cultures have tobacco as an excitant, and you have a handful of maybe three or four more. And that is the substance of plants that were used to affect personality. And that went through the 1200s, 1500s, 1800s. You're up to the beginning of the nineteenth century, you still had this. You knew microbes existed, but you didn't know they associated themselves with disease.

At this point you had the very first discovery—I will get into some of these as I talk about the individual drugs, so I want to keep it now to more of a global sense—of chemotherapy where you can take a chemical and design it to do a specific thing in the body. The concept of chemotherapy is less than a hundred years old. All the organic chemistry had its nucleus in the early 1800s and was really quite a rolling sophistication in 1860, 1880. There was no concept that these things were drugs or could be used as drugs. There was no concept of drug! The causes of illness were in the stars, so to speak, not in the microbes. They were due to this bad relationship with that. It was due to having done something or been somewhere, or not having said your something or other, or having associated with something else. It was known during the Black Plague that where rats went the plague followed. No knowledge that the rats brought the plague. There was no association of a disease with a tangible something you could get at. Diseases were God given. God gave diseases, God took them away. And you prayed and you read incantations and you went to the shaman, medicine man/woman, whatever, to get over your disease. It was not until the beginning of this century, really, that the whole concept of an infectious organism came to be known.

[Directed to student] Yes?

STUDENT: Even so, wasn't there a really big struggle in the medical profession in the earlier part of this century to have doctors wash their hands?

SASHA: It was invoked in the mid-1800s and some said that's fine, but it has nothing to do with medicine. Oh, yes. Cleanliness was very much promoted. Other people promoted dirtiness because it was a badge of effectiveness and the promotion of wrapping the bloody rag around the telephone pole became the barbershop. Display the blood. Show that you're effective.

STUDENT: I remember hearing there was a real problem with when a woman would go into the hospital, and the doctor had just cut off someone's gangrenous leg, you know, and then delivered a baby, and the woman would die. And they didn't understand why. The midwives had a much more effective rate of survival, but they still didn't understand that. Then some doctors said we should wash our hands, and there was a huge controversy.

SASHA: Yes, this preceded the knowledge of why by some time. But it was not universally practiced. It's not practiced today in many areas. So yes, you have the knowledge, and it was done by some, but the cleanliness by trial and error proved to be more effective. But remember, each person who's ever dealt with illness—this is one of the basic things in medicine—deals one on one and is rarely in the position to generalize.

This is the basic difference between teaching the concepts of medicine to people in an academic institution for the PhD degree, where you are a doctor of anatomy or a doctor of pharmacology or a doctor of this or that science. There you are trying to find things that tie all these together, like looking at cleanliness and a better recovery rate in the hands of a midwife. You're trying to generalize all these together and if you then can apply it to a specific case, you are considered quite inventive and quite creative. The concept in medicine, and this has been true for millennia, is to treat the individual, and if you can tie it into a generality, then you are inventive and creative. The whole philosophy of science and medicine are really at opposite ends of a teeter totter and it's very few who can see both ends of this: the concept of generalities and then applying specifics, or the concept of specifics and trying to find the generalities. Most of the people in medicine were after the specific. This person survived. They didn't connect the fact that the cleanliness had given them better statistical average because they were focused on each individual case. But this does work into the unconscious of the psyche in time and it's one of the things that led to the uncovering of the origin of infection.

The very first understanding of an alkaloid of a plant being other than something like an opium, where the plant was a homogeneous something, and you ate the opium and it did something, the isolation from the opium of morphine, from the opium poppy, was not achieved until the beginning of the 1800s. It was the first time they realized that there were materials in nature that were not plants that had actions.

The synthetic chemistry of 1870, 1880, 1890 was quite, quite elaborate, but not directed toward drugs. The word pharmacology was created in 1890. It never existed before that, the idea of drug action in the body. Everything before that in medical school was *materia medica*, things that came from nature that had medical action. But not the idea of drugs. The first real direction of that was—oh, who was it who created salvarsan, the arsenic compound against syphillis? I think it was Lister. No, I'm not sure who it was.[3] But it was the idea of tailoring a molecule, the idea of the magic bullet.

STUDENT: Was it Pasteur?

SASHA: Not Pasteur. Aw, gee. But it was the idea of going into the laboratory, making bumps on a molecule, making it wide, making it narrow, high, short, modifying the molecule, seeing how it affected the microbe that was at this point associated with syphilis, and how it affected the person who had the microbe. The idea was to kill the microbe without killing the person. And the fractionation of toxicity was the first really directed action toward, as we call it, chemotherapy. And the attack on syphilis was the first success, carrying partially into the microbe, into the spirochete, without killing the host who carried the spirochete.

The first evolution of what we know as antibiotics really got its start in about 1920, 1930. It was the development of sulfanilamides. This was again in Germany, which was the heart of chemistry up until World War II. Organic chemistry was, in essence, a German science. Prontosil, I think, was the dye that was involved. There were certain azo dyes that seemed to make animals that were sick get better. They had no idea why. They argued, "Well, this color dye would do it. That color dye would not." But it turned out, it was only one or two specific types of dyes that would do it. And they took this dye into the petri dish where the organism grew, and the organism kept growing. The dye did not do it. It turned out it was the dye in the animal that did it. The animal metabolized the dye

---

3 This was first synthesized by Alfred Bertheim while working in Paul Ehrlich's lab.

to a component that was sulfanilamide, and that was the thing that killed the organism. Then they found that by making different substitutions on sulfanilamide a whole area of health treatment by drugs came to be. In 1930, 1940 the life-saving drugs were the sulfanilamides.

Back in the 1930s was the first discovery that some microbial products can kill bacteria.[4] And by 1940, the evolution of penicillin came on as being a microbial toxin that was not toxic to the microbe that made it, but was toxic to microbes that were in that area. And penicillin became the first of the antibiotics about the time of World War II. The development of other antimicrobials from other bacteria from 1950 to 1960 was really the age of antibiotics. And now with the exquisite skill of being able to put the molecules together in the most abstruse ways, you have really the development of chemotherapy as being a major art form.

But the whole concept of the use of drugs really did not exist until about a hundred years ago. Before that, it was plants that affected the organism itself.

STUDENT: When did they start being outlawed?

SASHA: Be what?

STUDENT: Outlawed.

SASHA: Oh! Good! Thank you! I wanted someone to ask about laws because I want to get into the history of laws, which parallels this whole thing in exactly the same way. The first laws against drugs came about around the turn of the century or just before that.

Every plant that has come into a new culture at some level or another has been rejected by that culture. It has been used by a few and rejected by most. When tobacco came into Europe, being brought back from the New World by the first explorers from Spain and Portugal, it was considered to be a truly evil thing. Tobacco was actually outlawed in several countries and, for short periods of time, had a death penalty

---

4 The organism that makes penicillin is a mold, a fungus, not a bacterium.

associated with its use. The concept of smoking was not in the Bible and was absolutely contrary to any teaching. In fact, the whole concept of smoking is a New World contribution to the use of drugs. There was no individualized smoking of any drug in the Old World, the European or Asian area, until it was introduced from the Native Americans. They smoked their tobacco and it's marvelous to read how they smoked it. Those who now have seen people smoking with the ease of the cigarette to the mouth must realize the original smoking was done through the nose. The Native Americans would wrap the tobacco and other plants into a tube and stick it up the nostril and smoke that way. The idea of going through the mouth was not ethical in their culture. But it became part of the culture when it was taken back to Europe, the mouth was used, and then it swept on around the world. The smoking of opium, which is a major mechanism of opium administration, in Asia, followed the smoking of tobacco in Europe, which followed the smoking moving from the New World to Europe. Smoking was unknown outside of the Americas prior to about 1500.

Coffee, the same thing. In fact, the textbook, *Chocolate to Morphine*, does delve into smoking and coffee. They're both paneled with a fair amount of detail. The "Coffee Cantata," I think, is discussed with some elaboration on Bach having written this little, marvelous cantata about a girl who wanted to marry someone she met in a coffeeshop and the father thought the coffeeshop was immoral and forbade her to have coffee. That's a quaint little story, but at that time coffee was considered, also, very, very satanic and a plant that achieved no good.

The primary plant that really invoked the structure of the law was opium. Morphine itself was very heavily used in the Civil War in the early 1860s. At this point, the hypodermic needle, the so-called hollow needle, had been invented in Europe, and the syringe that pushed the thing through that needle had been developed about 1850 and was really instrumental in the Civil War in the treatment of pain. And the hypodermic syringe and hypodermic needle was used at that time. Infections must have been something but remember that the body has remarkable resilience against infections. The pain was the thing that was intractable in wounds and injuries; and pain was treated in this way.

At the end of the Civil War, there was a broad use of morphine and opium, and it was a major drug in this country up to the turn of the century. This was aggravated by the bringing in of very cheap labor from the Orient. The Chinese were brought in to work on the Transcontinental Railroad and brought with them the use of opium. In fact, I'm going to get into the structure of the Opium Wars and how opium got into China in the beginning of the last century in a whole lecture just on morphine and opium, so I don't want to belabor the origins of that. But the use of opium was quite broad, the use of heroin was very broad in this country at the turn of the century. And this is the first drug that an effort was made to control.

[Directed to student] Did you have a question?

STUDENT: Yes. Is it true that the clergy in China used to push the people to smoke opium?

SASHA: It's tricky to say the clergy did. Really, it was the British. The British wanted to open China to trade, and the way they did it was to bring in opium and force the Chinese to buy opium. Brought it from India and from Ceylon and from the lower areas. Where is the origin of opium? The opium poppy originated in Europe. They found fossils of opium plants in old fossil beds in Switzerland, in that area. And they believe the origin of opium was in Europe, and it traveled east, just as the origin of hemp was in China and traveled west to cross over the centuries in the southern part of Asia Minor. No, it was largely the British. The whole story of the Opium Wars was in the 1820s and 1830s. It was the British wanting to open up China to trade and force the opium in.

The Chinese brought opium to this country [the United States]. And once the Chinese were no longer wanted, and they became presented as an undesired minority, the whole anti-opium mark really became quite strong—the opium den, the destruction of culture by means of opium. Meanwhile, they're selling over-the-counter carpet bagger medicines to cure what was called the "woman's disease." There were probably three times as many women than men associated with opium use because it was used for all complaints, everything from amenorrhea to headache to misbehaved children. They were given morphine in one form or the other. And there was an effort to stop the importation because it was out of control. They passed a tariff against it. It was smuggled in anyway, a lot of it coming in from Canada. Smoking opium was separated from non-smoking opium or eating opium. And the first serious law came in 1914. This was known as the Harrison Act. It was passed in the beginning of World War I and it established the flavor of the laws against drugs that persisted for fifty-five years up until 1970.

What the Harrison Act did was to make the control of drugs a financial thing, not a criminal thing. The Harrison law was regulated by the Department of the Treasury and it was a law that said: You can bring in opium, you can bring in heroin, you can sell it (heroin at that time was quite broadly used). You can bring these things in, but you must pay a tax. You must have a license. I actually dug out, from files at home, a quotation from that law that really shows the way the flavor went. Here we go. In the law it says most over-the-counter preparations are exempt as long as they contain less than two grains of opium. By the way, this is a term that you're going to find in the old literature: "grains." A grain is a weight, and, say it's around sixty milligrams. A typical dosage of morphine would be five or ten milligrams, maybe an eighth or a fourth of a grain. You could bring in two grains of opium, a quarter grain of morphine, an eighth grain of heroin, but here's the point: the physician's rights. Remember, there has always been a lip service given that laws will not get in the way of the practice of medicine. So, at this point, they actually wrote into the law the following: "Nothing contained in this act shall apply to the dispensing or distribution of any of these drugs by a physician in the course of his professional practice only." Saying, in essence, "Physician, you do what you want, we're not going to get in the way of your professional practice."

That was the terminology that got distorted in about two years because what was said was, the people who use heroin, the people who are dependent upon (addicted was the common term then), addicted to, heroin or morphine were criminals because they hadn't paid their dues, hadn't paid their taxes. They didn't have a license. They were criminals. And no physician who was pursuing professional acts can deal with criminals. He can only deal with patients. These are not patients, they're criminals. And so they said to the physician, "You can't treat a heroin addict because they are not a patient, they are a criminal. We gotta put them in jail." The physician said, "But, they're ill, they're sick." "It's not a professional thing." The physicians were hurt, they were badly hurt. By hurt, I mean about a half a dozen of them were criminally charged and put in prison because they treated heroin addicts as patients. The law said, that is not a professional act. The whole criminality of addiction was instilled about 1920 by this move.

And this development went on over time, about the same time the Pure Food and Drug Act was created for the sake of labeling. That was created in the Department of Agriculture. Now you say, well of course it should be under Health, and it is. The fiscal laws that deal with the regulation of drugs should be in Justice, and now they are, but at that time it was Treasury and Agriculture. This was primarily heroin, and a secondarily major problem that was believed to be a serious one at that time, was cocaine. I would say the amount of heroin and the amount of cocaine are not less today than they were then. And if you consider per capita, they're both probably larger today than they were then. What is the heroin usage in this country? A million, or millions? What is the cocaine usage in this country? Millions, I think, without any question. In fact, the latest issue of the annual publication from the Department of Justice on the drug use in this country has been delayed again, has not appeared, because no one can agree on how much cocaine is being used. They just don't know how much cocaine is used. But millions is a fair guess.

At that time there was a lot of marijuana, but no one perceived marijuana as being a drug. It was something that was used by people who didn't count. And it was not really considered until about 1920. A name that was very famous then, Harry Anslinger, a very, very ambitious person, decided to make marijuana a device by which he could himself achieve a certain amount of authority. He championed the passage of the Marijuana Tax Act and the development of the first of the government agencies, the Federal Bureau of Narcotics, which was created in 1930 and was made in a large measure to try to control heroin, opium, and marijuana. Marijuana was promoted at that time as being a true killer. I think everyone has at one time or another seen some of the campy movies and the posters of the period, "The Killer Weed," and I don't know what all. "The Weed That Descends You Into Hell" and what have you. It was known, and every single thing supported the fact that marijuana was the gateway to crime and was the gateway to more drug use. And it was brought under, again, a financial control with the Marijuana Tax Act, not illegalization, but the requirement of a taxation.

Really this went on through World War II. The real change occurred in this whole philosophy about 1960 when it was found that marijuana usage was very broad. This was 1960, the time of Haight-Ashbury, the time of the flower children, the whole generation that was totally oriented toward drugs, toward self-appreciation and contentment. Drug use was very broad. There was a great deal of offense given to the authorities by general drug use. An attempt was made to separate those drugs that had medical utility, but were potentially abusable, from drugs that were simply abusable. And there was a move of all the drugs that were stimulants, depressants, hallucinogens, which were transferred from the Federal Bureau of Narcotics, the FBN, into the FDA under a division that was known as BDAC [Bureau of Drug Abuse Control].

STUDENT: What year is this?

SASHA: About 1965. This became BDAC, Bureau of Drug Abuse Control, under the FDA. This is really kind of the beginning of the chaos that is now, still, in its own way, going on. At that time, you had the Narcs, who were the Bureau of Narcotics, and they were after cocaine and heroin. Cocaine, although you know it as a stimulant, and medically it's known as a stimulant, and the textbooks say it's a stimulant, in the eyes of the lawmakers it is a narcotic. So cocaine and heroin were narcotics, marijuana was a hallucinogenic, and speed was a stimulant, and barbs (barbiturates) were depressants and they're all in the FDA under BDAC. There were what were called BDAC agents. They were like Narc agents except when they went on a raid, they wore green bandanas instead of red bandanas so that the Customs agents could wear purple bandanas and the people from the FBI would wear another color, yellow. And so, you would know you don't shoot at anyone in a yellow bandana if you have a red one. It was a weird, weird collection of authorities. And the BDAC was suddenly given the part of the FDA: "Is this teratogenic? Does this cause cancer? Is this safe to deliver?" Suddenly they had guns, badges, and were out there as policemen. No experience in this at all. And it was total chaos for about three years until it was finally dissolved. The BDAC group was disbanded. The FBN was disbanded and the two merged into what was called the Bureau of Narcotics and Dangerous Drugs. This was in, roughly, 1968.

STUDENT: What does BDAC stand for?

SASHA: Bureau of Drug Abuse Control. And the Bureau of Narcotics and Dangerous Drugs. In 1968, this is the move that is going to really put this thing under control. It went on for about two years and clearly it was not working. And what happened was that then there was a complete restructuring of the drug law. 1970 was the passage of the Controlled Substance Act, which, in essence, ended the Harrison Narcotics Act. And that is the law that is in effect today.

In this case, drugs were given priorities as to how bad they were. You had the development of five schedules of drugs, Schedule I being the worst and the least useful and the least nice. Schedule V being the least offensive and the most useful. Schedule I drugs were drugs that had no medical utility and had a high abuse potential. Schedule II were drugs that had medical utility, but had a high abuse potential. III was less abuse potential than II. IV was less abuse potential than III. V was less abuse potential than IV. So, you had a sort of dropping of abuse potential.

Nowhere do they say what abuse potential is, nowhere do they say how do you determine this has four point seventy-two times as much abuse potential as that. These numbers cannot be made. But nonetheless, the flavor has to be big abuse potential, no medical utility continuously over to maximum usefulness, minimum abuse potential, minimum controls. Schedule I and Schedule II are the ones that sit over here as being the ones that have drama, and you'll find all your dramatic drugs are in there. And Schedule III, IV, and V are things like sleeping pills and more minor tranquilizers. So that category exists today, the scheduling of drugs. And that is an outgrowth of the Controlled Substance Act of 1970.

THE FIVE SCHEDULES OF THE CONTROLLED SUBSTANCES ACT

SCHEDULE	ABUSE POTENTIAL	ACCEPTED MEDICAL USE?	RISK	
I	High	No	There is no accepted safety under medical supervision	
II	High	Yes	What is the dependency risk that will follow abuse?	
			Psychological:	Physical:
			Severe	Severe
III	Less than with I & II	Yes	High	Low or Moderate
IV	Less than III	Yes	Limited (Less than III)	Limited (Less than III)
V	Less than IV	Yes	Limited (Less than IV)	Limited (Less than IV)

The procedure needed to place a new drug within this structure involves a number of steps. An announcement of intent must be published in the Federal Register. A waiting period of 60 days allows comments and objections to be received. If it is felt to be appropriate, hearings will be held to address these matters. Otherwise, a final action can be taken at the sixty day period. The recently passed Comprehensive Crime Control Act of 1984 allows the bypassing of this delay in instances of drugs that are felt to present an especially serious threat to public health. A temporary emergency scheduling may be invoked for a period of one year, to apply during the period allowed for comments and hearings, and it can be renewed for an additional 6 months following that first year. Action must be taken before the expiration of the renewal period.

At that point, the big change that occurred after that was probably the one now in existence. In 1972, the Bureau of Narcotics and Dangerous Drugs was dissolved and, this was under Nixon, the Drug Enforcement Administration (DEA) was created, which is in effect today, although it is losing control of its own offices. I'll talk about this in a minute. And this was a transfer, finally, of the drug enforcement concept from Treasury to Justice. And the DEA now answers to Justice. So now, the drug acts are really a crime against law rather than a financial manipulation against taxation and licensing.

From about 1972, '73, '74 on, there has been approximately a major drug law every year. And I've jotted some of these down just to give you the flavor of what's been going on in the last ten or fifteen years. Everything is geared toward more penalty, more restriction. The Boggs Act says that you may now take some discretion away from a judge as to whether they can make a conditional sentence. I'll give some specifics. The Psychotropic Act of 1978 for the first time invoked criminal forfeiture as a penalty for drug crimes, for which you have been committed. You may take not only some of a person's life away and so many years in prison and so much fine and so many dollars, but you may take their land, you may take their car, you may take whatever you wish as a penalty for having committed a crime. Criminal forfeiture was spoken against, but in the laying down of the laws by the founding forefathers in 1770 and 1780, this is the first allowance of criminal forfeiture in our legal system.

In 1981, the Senate Drug Enforcement Caucus led to the Department of Defense Authorization Act of 1982. I can give you copies of this if anybody wants them. It was the first time that allowed military involvement in civil law. This was the first federal act that said the military can be authorized to pursue civil law enforcement. It allowed the Coast Guard, the Army, the Navy, Marines, for all I know, to go out and actually perform acts that led to civil law enforcement. That was authorized in 1982. It did not exist prior to that.

Another law, the Tax Equity and Fiscal Responsibility Law of 1982, removed restrictions on tax records if they dealt with the prosecution of drug cases. So for the first time, income tax records and payment tax records that are normally privy between you and the tax collecting agency can be brought into a criminal case dealing with drugs. In 1984 the Comprehensive Crime Control Act was, besides revising penalties and making them more severe, the first to invoke emergency scheduling. And here I'd like to go into a little bit of how drugs are put under the law.

When they made The Controlled Substances Act in 1970, they took all the drugs that were known to the DEA (at that time, the Bureau of Narcotics and Dangerous Drugs) and all the drugs that the FDA had been pursuing, and wrote them into the law. These are the drugs in Scheduled I through V that are considered illegal, and the possession of or the sale of, importation, exportation, manufacture of, will be a crime unless it's authorized. Now, the laws were extremely explicit. They said, this, this, this, and this. There were 270 drugs in the law and that was it. Sometimes there are extensions, such as "all salts," or "all derivatives." They never defined what derivatives were. "All isomers" they only defined as being optical isomers, they didn't think there were positional and structural as well. And these were spelled out as being the things that are controlled. If it's a barb, it should be a barb that causes a depressive effect on the nervous system. Other barbs are not controlled.

Well, the thing is you have 258 drugs, here comes drug number 259. You write it into the law, but to do that you have to make a public announcement. There is a publication that comes out every day of the week from the federal government called the Federal Register. And in that Federal Register all things that are written as regulations or things that will change the law are written down explicitly to make a public record. If the IRS says, "We will now require a thirty day something or other before something or other," it's written in the Federal Register. That is the voice of the government to the people. This is how you know you cannot import oranges after September as of such and such a year without paying a fine. You want to make a drug illegal, you put it in the Federal Register and wait sixty days. During that sixty days, hearings can be requested, people can object, people can say, "I think it's a good move." At the end of sixty days, then you can change the law to invoke this new drug into the regulations of the law.

Now, the Emergency Scheduling Act of 1984 allowed drugs to be put on there without this hearing period, put on within thirty days, but with no complaints for a year, and during that year then we will have discussions and hold hearings if necessary.

[Directed to student] Yes?

STUDENT: How about after that year, like MDMA?

SASHA: The law in the Emergency Scheduling Act is set for one year. It can be renewable for up to six months.

STUDENT: So has it been renewed?

SASHA: It has been renewed and it's been made permanent.

STUDENT: As Schedule I?

SASHA: As Schedule I. So it's no longer illegal by Emergency Scheduling, it's illegal by proper procedure. This was a way to allow drugs to be put on fairly quickly without going through the time-consuming and potentially objectionable delaying procedures of information. Then the one that was just now enacted was called the—I have down here—the Controlled Substance Analogues Enforcement Act—it has a little bit more general name, but it was passed just a few months ago in November of 1986. I consider it to be one of the most freedom restricting things that has ever been put into the law. It makes any drug that anyone wishes to make, illegal. If it meets certain very restrictive requirements, chargeable as if it were illegal. It does not make things illegal that were not recognized as illegal, but makes them punishable as if they were. Its common name is the Designer Drug Bill. What it says is that any drug, any chemical, is illegal if it meets the requirement of being substantially similar in structure to a Schedule I or II drug. With Schedule I or II drugs you can start looking at morphine and narcotics and mescaline and a host of structures, and I think you could argue any drug, as having some similarity. If nothing else, it has carbon, hydrogen, oxygen, and nitrogen. It has some similarity to a Schedule I or II drug. So that is not the way that you can restrict things being or not being chargeable. But rather, if it contains a stimulant, depressant, or hallucinogenic action. What this does, in essence, is to erase all uppers, downers, or stars.

That whole concept of ups, downs, and stars is quite common. There is a three-ring binder that is put out by the federal government that has on the binding the following: it has arrows pointing up, it has arrows pointing down, and it has a bunch of stars. Apparently, the federal government in its making spectra and recipes and methods of finding drugs, and what they had available—it's a treasure book of spectra, of recipes, what have you, that's restricted for government use. They didn't know how to name it because they didn't know what kind of a term would embrace all of these. So they put on the binder arrows pointing up, arrows pointing down, and a bunch of stars, presumably for

stimulants, depressants, and hallucinogenics, which I think is a nice, almost cuneiform basis for identification. [Laughter.] And it's known in the trade as "Ups, Downs, and Stars." That's the name of the book. And you go to the federal publishing house and "Is there a new issue of 'Ups, Downs, and Stars?'" You get "No, it won't come out until—" That kind of a thing.

These are the three properties. If a drug has any of those properties and is intended to be administered to humans, then the possession of that drug, the giving of it, the buying of it, the selling of it—not the using of it, that's still not in the

Analysis of drugs

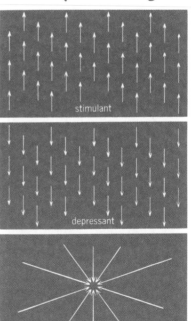

stimulant

depressant

hallucinogenic

general law—the importation of it, the manufacture of it, the exportation of it, is a crime that can be punished as if it were a scheduled drug. This law is in force, but it cannot be gotten, it's not going to be available until the middle of the month, although technically it was signed into law on November 7.

[Directed to student] Yes.

STUDENT: Is it being determined by each court case?

SASHA: Has never come to a court case yet. This is still untested, but it will be. For example, take the term—god, we're running out of time!—take the term "substantially similar." Most people who have tangled around with the aspects of rhetoric know what's meant by a disclaimer. It's when you put a word in there that gets you a little bit oozy out of it. You don't have to go through with it if you don't want to. Like "almost" or "about" or "by and large" or "pretty much." I worked for a while with a person who got three of them into one sentence which I think is a world's record. He said, "We will probably ship (probably: number one) the sample in seven to ten days (number two: range) or two weeks at the latest (number three)." I mean, these are disclaimers.

So you say something is "similar to," that's a disclaimer. Something is "substantially the same as," that's a disclaimer. What is meant by "substantially similar?" You have two disclaimers, and you have a feeling that it's put in there to give enough latitude that you can use it the way you want. Is the taillight structure of a 1986 Pontiac "substantially similar" to the taillight structure of a 1984 Chevrolet? Well, in some ways yes and some ways no. Is this drug structure "substantially similar" to that? Some ways yes, in some ways no. You're going to have a conflict of expert testimony, "They're totally different!" "They're substantially the same." And when you get experts on a stand, each holding forth under oath, you want to have a balance of who the jury believes. The jury happens to like this expert's color sense more than that expert's color sense. Have you ever seen experts on the stand? The jury responds to the experts—I've been an expert on the stand, I know—you get a person on that stand, cross the legs to protect the genitals, cross the arms to protect the breasts, put something in your mouth, this is the body language that says "lie." No matter what is said, it is a lie. This is the body language that says "truth." [Gesturing. Class laughs.] This is what juries respond to and they respond to the nature of the expert, not what they say, but because experts say, "That person was clearly insane at the time he pulled the trigger!" Or, "That person clearly knew what he was doing at the time." You want to know, was the person guilty of a crime or were they off their rocker when they shot someone? Experts will say, "This drug is similar to that" as far as its structure goes. If it has this or this or this action, and it is intended for humans, it is a crime, a felony, to have that drug unless it has been approved by the FDA, which, in essence, puts the FDA in the position of approving all human research.

That is written in the law. I consider it to be a very, very, very destructive move, because all of the drugs that we have stemmed from things that have been found by putting them into human beings. I'll give you the whole story sometime of how they discovered nitrous oxide and ether and morphine and all of these. They were done by gathering around a table, "Joe, you take one grain; Mike, you take two grains; Mary, you take four grains; and we'll see where the anesthetic level is." [Laughter.] That's how drugs were found. You find drugs that are psychotropic, that change the state of mind, that make you happy, make you sad, make you insightful. You don't find that action in mice and guinea pigs. You find it in humans. And that is now illegal. It's a very tricky law and it's the latest one. However, tune in again this same time next year and we'll find what new laws—

Anyway, I want to get into plumbing next hour, plumbing of the human body, and it's going to be kind of a fun one in its own right.

Note: Sasha's original Lecture 3 notes, typewritten in 1987, will appear on pages 301-318, following the end of Lecture 8.

Include as part of Lecture 3

THE HISTORY OF THE FEDERAL DRUG LAW

The history of man has been told with many different frameworks as guides. Psychoactive drugs can serve as one of these. Man has historically been, and still is, associated with the use of drugs that are able to modify his body or mind. Each person has inevitably found one or more that he personally approves of, and believes that others are inappropriate. To him, involvement with the first group constitutes drug use, and with the second, drug abuse.

Records of use of some drugs have existed from pre-Christian times: there is the use of Opium (poppy seed eating in mid-Europe, 2500 BC), of Alcohol (breweries in Egypt 3500 BC) and of Marihuana (China, 2500 BC). The earliest use of others may be equally ancient but the origins are more difficult to document (Khat in the Near East, tobacco in the New World, Peyote and Psilocybe in North America, and dozens of other natural toxins in hundreds of independent cultures). The American obsession with drugs use and abuse springs from a rich universal heritage of social approval and disapproval of drugs. In the United States, the rebellions against the British tax on tea (Daughters of Liberty, 1770; Boston Tea Party, 1773) quite literally led in sequence to the passage of the English Parliamentary Coercive Acts, the formation of the First Continental Congress, and to the War of Independence. All of the earliest efforts to regulate and limit the use of drugs were enforced through the fiscal device of taxation, an approach that had remained in use until the passage of the present Controlled Substance Act (1970).

The record of the laws concerning Alcohol can be used to give an interesting historical perspective to our present day efforts to contain drugs such as Heroin and Cocaine. The medical profession has long held that the overuse of Alcohol is a physical disease, but the many temperance societies (the first was formed in 1789, and there were over 6000 by 1833) held that any Alcohol use implied moral incompleteness. The medical and moral agreed that restrictions were needed. Some local legal controls were instituted as early as 1845 and a move was made on the national scale in 1862 with the passage of the Internal Revenue Act (a tax on all retail liquor). The political picture reflected this consensus with the formation of the Prohibition Party in 1869 and the WCTU in 1874. Between 1880 and 1900, the inclusion of "temperance education" and "anti-alcohol teaching" was made a legal requirement in all states. Yet, during this period from 1870 to 1915 and the passage of the Sixteenth Amendment establishing the authority for the Federal income tax, between one half and two-thirds of all the internal revenue of the United States came from the taxation of alcohol. The culmination of this anti-alcohol hysteria was the passage of the

1

Eighteenth Amendment (prohibition) in 1919 which was not repealed until 1933.

A similar story can be told with cocaine, with heroin, with marijuana, except that in these examples there has not yet been any move seriously entertained that would be parallel to the repeal of the laws that prohibited transactions related to Alcohol.

The background leading up to the passage of the Controlled Substances Act in 1970, and the changes in it, and amendments to it, are listed below chronologically. Several of the recent laws had been passed with some purpose other than narcotic law enforcement as an intended goal, but they are included since they have been found to be valuable when applied in this direction.

HISTORY OF DRUG LAWS:

1887 (February 23, 1887) One of the first laws of the United States against drug abuse prohibited the importation of Opium into the United States by any Subject of China (this act could be punished by from one to six months in prison or up to a $500 fine).
     Although opium could be legally imported from Canada through the 1890's, there was concern over its importation by smuggling. When the duty was lowered from $12 to $6 a year, the smuggling stopped, and the volume of legal importation soared to 100,000 pounds per year.

1903 (August) There was established a U.S. Opium Commission to study the methods of regulation and control of Opium. During the years of this study, the use increased and according to the Federal Opium Commissioner, Hamilton Wright, some 160,000 pounds of Opium were imported for smoking and eating in 1907. And in 1906, 2,600,000 pounds of Coca leaves were imported. These abuses led to the Harrison Narcotics Act (1914-1915) which was the principal drug law of the United States for 55 years.

1906 Pure Food and Drug Act was enacted, with the purpose of preventing the writing of false and misleading labels. This was not an antidrug law but rather an indirect attempt to regulate drug distribution. The act was ineffective as it was essentially never enforced.

1909 Congress enacted the first legislation aimed at controlling drug abuse, the Act of 9 February, 1909, which specifically prohibited the importation of Opium for other than medicinal purposes. That year there was

2

a Conference of the International Commission in Shanghai which recommended that all participating governments (thirteen, including the United States) take drastic measures to control the manufacture, sale and distribution of morphine.

1912   The Hague Convention was held, and concluded that there was a need for the control of the domestic sale and use of opium and of coca.

1914 (December 17, 1914) President Wilson signed 26 U.S.C. 4701, the Federal Narcotics-Internal Revenue Regulations, commonly called the Harrison Narcotics Act. The effective date was March 1, 1915. This Act divided drugs into four classes: Class A contained drugs that were highly addictive, Class B contained drugs that were considered to possess little addiction liability. Classes X and M contained exempt and especially exempt narcotics, respectively. This bill was a tax measure, not a commerce measure, and was written to license and tax all who import (or manufacture, sell, or dispense) opium or cocaine (and their derivatives).
        A companion act was passed the same year, Narcotic Drugs Import and Export Act. Neither of these bills received much comment at the time -- the intense debates were concentrated on the prohibition movement.
        It is interesting that at the time of the signing of this act, cocaine was considered as a much more serious problem than either heroin or the opiates. At that time 46 of the 48 states had already adopted legislation regulating the distribution of cocaine; only 29 had any legislation involving opiates.
        A problem that developed at this time was the assignment of addicts (by physicians) as patients. The Act was amended (Act of February 24, 1919) to stop the maintainence of addicts by medical treatment. Thus the law enforcement community viewed users as criminals. Physicians were arrested and convicted for their stand on this matter.

1922   An amendment to the Narcotic Drugs Import and Export Act of 1914 banned cocaine entirely, and placed severe restrictions on the importation of coca leaves and Opium. This was the first definition of cocaine as a "narcotic."

1929 (January 19, 1929) President Coolidge signed the Porter Bill which established two narcotics farms for the confinement and treatment of federal prisoners who were addicts. This was an acknowledgement of the physician's difficult relationship in regard to the Harrison Act. On May 25, 1935, the United States Narcotic Farm at Lexington, Ky. was dedicated, and it was designed to treat 1400 addicts (voluntary patients and convicted violators of federal law).

3

1930 (June 14, 1930) Congress established the Bureau of Narcotics, within the Department of Treasury. President Hoover appointed Harry J. Anslinger to head the Bureau on September 23, 1930. He initiated the policy of attacking the supply of narcotics at its source.

1933   An international treaty was ratified by the League of Nations, limiting the manufacture of narcotic drugs to legitimate needs. This was hailed as a means of ending narcotic addiction.

1937 (August 2, 1937) On this date the Marijuana Tax of 1937 was passed. In the early 1930's, the years of the depression, it became widely believed that Marijuana was addicting and that its use caused insanity. Anslinger spoke widely concerning its presumed role in crime and mental deterioration. It had not been considered in the Harrison Act, and there was felt to be the need of control. The 1937 Marijuana Tax Act was passed calling for up to a $20,000 fine and 20 years in jail if a person possessed marijuana without a transfer or possession stamp in his possession. The enforcement of this law was mainly directed against immigrants from Mexico. All of these moves and acts associated with enforcement were Treasury Department actions, rather than actions of the Justice Department.

1938   Food, Drug, and Cosmetic Act was enacted, requiring that drug manufacturers provide directions for drug use and proof of safety. The act was amended in 1962 to require the demonstration of efficacy, as well.

1942 (December 11, 1942) This is the date of the passage of the Opium Poppy Control. Again, this law was a revenue action, as its enforcement was directed towards the presence of a required tax stamp. The production of Opium had been encouraged domestically as it was difficult to import from oversees under wartime conditions. This law was an attempt to interfere with any illicit production.

1946 (March 8, 1946) The Narcotics Act of 1946 was passed as a parallel to the above Poppy Control Act, and provided controls for the synthetic equivalents of opium and its derivatives, and for other drugs such as cocaine.

1951   The passage of the Durham-Humphrey Amendment to the Federal Food, Drug, and Cosmetic Act, which required that drugs which cannot be safely used without medical supervision should only be dispensed by prescription. Both barbiturates and amphetamines were gathered within this amendment.

1951   The first of several amendments to both the Harrison

4

Narcotics Act and the Narcotic Drugs Import and Export Act, to increase penalties. Here the Acts were amended to standardize penalties between them. Here are the first requirements for mandatory prison sentences (they still could be suspended and probation could be given).

1954   A congressional hearing was held to evaluate the needs of imposing Federal controls on the distribution of amphetamines. No legislation resulted, but in the late 1950's the FDA began taking action against people who were diverting amphetamine from legitimate channels.

1956 (July 18, 1956) The Narcotic Control Act of 1956 was passed. It was felt that more severe penalties might work better. Not only were the prison terms and fines increased for the violation of the narcotics laws, but the minimum penalty limits became mandatory by the elimination of suspended sentences, probation, and parole.

1960 (April 22, 1960) The Narcotics Manufacturing Act of 1960 was yet another move made in the attempt to restrict and thus control the illicit use of drugs. Manufacturing licenses were required, and the Treasury Department could establish manufacturing quotas. This was a move directed towards the establishing of control of the synthetic drugs in a manner parallel to the existing controls directed towards the natural narcotics by the Import Export Act.

1961   The United States participated in the United Nations Conference directed towards the establishment of a single convention on Narcotic Drugs.

1962   A number of drug amendments were passed in 1962, but no action was taken on the attempts to include restrictions on amphetamine. In this year, a presidential commission recommended strong Federal efforts be made against the diversion, from legitimate trade channels, of amphetamine and other non-narcotic drugs.

1962   The Racketeer Influenced and Corrupt Organizations Act (RICO, 18 U.S.C. 1962, 1963) was enacted into law. The main purpose of RICO was the elimination of the infiltration of organized crime and racketeering into legitimate organizations operating in interstate commerce. There was no indication that this law had been specifically intended to address criminal activity such as drug trafficking.
        On October 19, 1970, Congress modified the RICO law to insert criminal forfeiture provisions.
        Although there is a rich history of civil forfeiture involved in proceedings against property which has been involved in some wrong, the act of criminal forfeiture involves seizure of property of a

5

person convicted of a felony. It was a common
punishment in historic England, but was specifically
prohibited in 1790 by the first congress of the United
States. As a result, criminal forfeitures were unknown
in the United States for 180 years. This amendment to
the RICO statute, and the continuing criminal
enterprise section of the Controlled Substances Act
(see below, 1970) are the first inclusion of this
penalty in American history.

This act has been recently applied to the Hells
Angels, in connection with methamphetamine traffic.

1963 (January 15, 1963) President Kennedy established an
Advisory Commission on Narcotic and Drug Abuse with
Judge E. Barrett Prettyman as chairman. The commission
filed its report at the end of the year with 25 strong
recommendations which prompted President Johnson (July
15, 1964) to direct that the full power of the Federal
government be applied to finding a solution to the
national problem of drug abuse. A major recommendation
was the transfer of all drug enforcement efforts to the
Department of Justice. This occurred in 1968.

1964   The Navajo Indians of the West Coast won their right to
use Peyote. Their religious ceremonies thus are not in
violation of the state narcotics laws.

1965 (January) Congressman Harris introduced the Drug Abuse
Control Amendments of 1965 (H.R. 2). This was approved
by congress July 15, 1965, and established within the
FDA, the Bureau of Drug Abuse Control (BDAC). It
became effective February 1, 1966, under the direction
of John H. Finlator.

The FDA had the responsibility for the enforcement
of this new law, which was limited in scope to
stimulants and depressants (the laws that related to
narcotics and marijuana were still administered by the
Bureau of Narcotics under the Department of Treasury).
For the first time, the authority of the FDA was not
restricted to interstate commerce in drugs, and its
agents could now, in some cases, make arrests without
warrants. Although the role of the FDA in this
particular matter was related to criminal law
enforcement rather than to medical issues, it still
engaged itself in educational projects geared to public
consumption. However, despite this reorganization, the
use of drugs continued to climb, and three years later,
this structure was changed again.

1968 (April 8) A presidential reorganization plan (Johnson)
merged the Bureau of Drug Abuse Control (of the Food
and Drug Administration, under the Department of Health
Education and Welfare) with the Bureau of Narcotics (of
the Treasury Department). The product was given a new
name, The Bureau of Narcotics and Dangerous Drugs.

6

This brought the entire drug abuse and law enforcement problem under a single authority, and under the Department of Justice.

1968 (October 24, 1968). A number of additional amendments, and modifications of amendments to the Harrison Act were enacted. These were largely directed to again increase the penalties for current law violations.

1970 (October 27, 1970) The Harrison Act is effectively removed from the books with the passage of the Controlled Substances Act of 1970. Title II of this Act was entitled Comprehensive Drug Abuse Prevention and Control Act. The synthetic psychoactive drugs, old and new, are now subjected to the same regulations as the older natural drugs. This is Public Law 91-513, dated October 27, 1970. It is offered as an amendment to the Public Health Service Act (and other laws) as a provision for increased research into, and prevention of, drug abuse and drug dependence; to provide for treatment and rehabilitation of drug abusers and drug dependent persons; and to strengthen existing law enforcement authority in the field of drug abuse.

With this act, congress effectively destroyed the Federal-State relationship that existed between the Harrison Act and the Uniform Narcotic Drug Act. To restore this, the Commissioners on Uniform State Laws drafted the Uniform Controlled Substances Act. This Act is structured on the Federal act (five Schedules, etc.) and requires persons involved in the manufacture, distribution, and dispensing of scheduled drugs to obtain a registration form from the state. Each state can, however, impose its own penalties for violation of these laws.

The (then) Bureau of Narcotics and Dangerous Drugs (BNDD) is the agency within the Justice Department delegated to implement the Act for the Attorney General, and its primary concern is diversion, distribution, and enforcement. There are several new provisions in the Controlled Substances Act. One is the shift from just sale or transfer of a drug, to its possession with intent to transfer it as also being a criminal act. Also, this law can be applied to acts that are committed outside the Unites States if they are performed knowingly or with the intention of eventually violating laws within the United States.

The Food and Drug Administration (FDA) is delegated to implement the Act with the concurrence of the National Institute of Drug Abuse (NIDA). The primary concern of the FDA is safe and effective medical use, whereas the primary role of the NIDA is scientific research.

Two aspects of this bill have received special attention through its enforcement in the last decade. Continuing Criminal Enterprise (CCE). A person is

7

defined as engaging in a continuing criminal enterprise if he is an organizer or supervisor of a group of five or more persons who have obtained substantial income through a series of violations of this Act. He is not only subject to more severe penalties, but he shall also forfeit both profits and properties associated with his felonious acts.

No-knock entry. Following the legal issuance of a search warrant, an authorized agent may enter, forcibly and without warning, if he feels that evidence might be otherwise disposed of, or if the warrant has allowed him to take such action.

See comments under the RICO act above (1962) concerning the history of criminal forfeiture.

1971   An international drug control treaty was written, the Convention on Psychotropic Substances, 1971. The United States became a party to this treaty in 1980, q.v.

1972 (March 21) Congress passed Public Law 92-225, the Drug Abuse Office and Treatment Act, aimed at reducing the incidence of drug abuse in the United States within the shortest period of time.

This established three new things. (1) Within the Executive Office, there was formed a Special Action Office for Drug Abuse Prevention (SAODAP). (2) There was established a National Council for Drug Abuse Prevention which provided for the formation of a National Drug Abuse Training Center. (3) There would be establishment of a National Institute on Drug Abuse (to be effective in December 31, 1974) with the eventual dissolution of SAODAP six months later.

It also authorized $1 billion to coordinate and expand Federal, State and private programs directed against drug use.

1973   The DEA was created. This organization was a merger of ODALE (the Office of Drug Abuse Law Inforcement, created in the Justice Department in 1972), of ONNI (the Office of National Narcotics Intelligence, also created in 1972 in the Justice Department), of the drug enforcement and intelligence functions of the U.S. Customs Service, and of the BNDD (the Bureau of Narcotics and Dangerous Drugs, which was itself a merger of BDAC (the Bureau of Drug Abuse Control of the FDA under the Department of HEW, and the Bureau of Narcotics, under the Department of the Treasury). The DEA is under the Department of Justice.

1974 (May 14, 1974) Nixon signed Pub. Law 93-281, the Narcotic Addict Treatment Act of 1974, which had the purpose of amending the Controlled Substance Act of 1970 to provide for the registration of practitioners conducting narcotic maintenance and detoxification

8

treatment programs for addicts.

1976 (August) The House of Representatives established the Select
Committee on Narcotics Abuse and Control to serve a
review and advice function over the DEA. Other
committees function in this area (the House and Senate
Judiciary Committees, and the Senate's Permanent
Subcommittee on Investigations) and finally in 1981,
the Senate Drug Enforcement Caucus was formed, to
support legislation wanted by the drug law enforcement
community. The DEA has provided the primary impetus
for most of the following laws and amendments.

1978 (November 10, 1978) Congress passed the Psychotropic
Substances Act of 1978 (Public Law 95-633). This is an
amendment to the Comprehensive Drug Abuse Prevention
and Control Act of 1970, and it allows the International Convention on Psychotropic Substances to
suggest additions to the Act. Approval from Congress
for this treaty had been sought since 1971. (The
United States became a party to this treaty on April
15, 1980, q.v.). In addition, increased penalties and
restrictions in the PCP area were included, including
the placement of piperidine in a position of having to
be reported when possibly associated with phencyclidine
manufacture.
    The very last section of this amendment is
entitled as "Forfeiture of Proceeds of Illegal Drug
Transactions." This addition states, quite simply,
that all proceeds from drug transactions may be seized
as forfeiture. This has given the government immediate
possession of boats, airplanes, real property, and bank
accounts, and has effectively limited the defense
capabilities of a number of defendants. This asset
seizure and forfeiture has been felt by many law
enforcement groups to have proven itself as an
effective weapon in the area of drug use prevention.

1980 (April 15, 1980) The United States became a party to the
International Convention on Psychotropic Substances,
1971, effective this date. Refer to the Psychotropic
Substances Control Act of 1978 for the groundwork
behind this move. All drugs that are recommended for
international control, are considered for scheduling
under the CSA of the United States. At the time of the
United State ratification of this treaty, there were
drugs listed therein which were not recognized by U.S.
law. The drug Parahexyl is an example, and reference
to 47 FR 33986 is illustrative of the processes that
have been set in motion.

1981   The Public Law 97-86 was enacted, entitled the
"Department of Defense Authorization Act of 1982."
This law includes a provision to revise the long-
standing Posse Comitatus statute. Prior to this law,

9

all military involvement in civil law enforcement was prohibited unless authorized by the Constitution or by an Act of Congress. With this law, there was a clarification of the role of the military in civilian enforcement activities, and the assistance and support services which may be rendered by the military to law enforcement officials are defined.

There was quick implementation in the formation of the President's Task Force South Florida in January, 1982. This operation was geared to the interdiction of narcotics being smuggled into Florida from the Caribbean and from Latin America. The military aid provided included complex logistic and vessel support, aviation and radar surveillance, and the loan of equipment and facilities.

1982 (January 21, 1982) Attorney General William French Smith announced his decision to give the Federal Bureau of Investigation and the Drug Enforcement Administration "concurrent jurisdiction" over violations of the Federal drug laws. The general supervision of this joint effort was put in the charge of Francis M. Mullen, Jr. The priorities of the relationship were defined by the AG's specific order that the DEA report to the FBI.

1982 (September 2, 1982) There was on this date enacted Public Law 97-248, the "Tax Equity and Fiscal Responsibility Act of 1982." Prior to this date the laws in effect (most recently, the Tax Reform Act of 1976) severely restricted disclosure of tax information. Complex financial transactions are often associated with sophisticated criminal activity, and this information was not available to Federal enforcement authorities. This new law included several provisions sought by the Justice Department "to facitiate the appropriate disclosure of tax information to Federal law enforcement agencies for criminal investigative purposes while maintaining safeguards needed to protect the privacy of innocent citizens."

Federal officials may now gain access to tax information which is a most valuable source of financial data necessary to prosecute narcotics trafficking and organized crime.

1983 The Comprehensive Crime Act of 1983 was presented to congress in its 98th session, as Senate S-829 and HR 2151. It carried some 42 reforms sought by the Department of Justice, involving bail, sentencing, civil forfeiture, the exclusionary rule, and drug regulation. This was passed as public law 98-743 below.

1984 (October 12, 1984) The Comprehensive Crime Control Act of 1984 (Pub. L. 98-47) was passed as an amendment to

10

section 201 of the CSA (21 U.S.C. 812). A House Report (98-975) that accompanied this law states (according to the Fed. Reg. Vol. 50 No.132, July 10, 1985) that:

"This new procedure (emergency scheduling) is intended by the committee to apply to what has been called "designer drugs," new chemical analogs or variations of existing controlled substances, or other new substances which have a psychedelic, stimulant or depressant effect and have high potential for abuse."

Chapter III of this law is entitled "Comprehensive Forfeiture Act of 1984." This portion extends the boundaries of seizure and forfeiture to include both terms "civil" and "criminal." This can allow seizure of profits derived from criminal acts as well as of property associated with the actual commission of such acts.

Chapter V of this law is in two parts: Part A (Controlled Substances Penalties Amendments Act of 1984) establishes weights and penalties within the law. Part B (Dangerous Drug Diversion Control Act of 1984) establishes or extends definitions of terms used in the law (such as isomer and narcotic drug), and allows the temporary placement of a substance into Schedule I, without the usual requirements, if this act is needed to avoid an imminent hazard to the public safety. In any given appication, this emergency scheduling will expire in one year unless extended for up to six months.

For the regulations established for the enforcement of this law, see 51 FR 5370.

1986 (October 17, 1986) A $1.7 billion anti-drug bill was passed by Congress and signed into law in the largest-ever financial commitment directed towards the drug problem. From the enforcement point of view it provides funds to Federal, State, and Local Law Enforcement groups, to the Coast Guard, to the Customs Service, for additional U.S. attorneys, marshalls, narcotics agents, and Federal prisons. Penalties are increased for nearly all convictions.

An amendment to the Controlled Substances Act was part of this, entitled The Designer Drug Enforcement Act of 1986 or The Scheduled Drug Analog Bill, originally introduced in 1985 as HR 2977. Although inspired by the appearance of several synthetic analogs of Fentanyl, in its final form it dictates that the laws that have been specifically written to pertain to Controlled substances shall now apply to any analog of these substances that is intended for human consumption. Exception is made only for drugs that have received FDA approval or exemption, so that in effect all research work with drugs known to have stimulant, depressant. or hallucinogenic properties

11

must be cleared through the FDA. Analogs are defined
as chemicals that possess a structure that is
substantially similar to that of a Schedule I or II
drug, and either have stimulant, depressant, or
hallucinogenic properties, or be promoted for human use
as having these properties.

12

February 10, 1987

Plumbing of the Human Body

SASHA: I want to start out and get as far as I can today in a very important aspect of drugs and drug action. That is, how they get around in the body. How you get them in, how they get to where they are going to do their thing, what they do when they get there (as far as one knows about physiology and anatomy without getting into the wet, gory details that I don't want to get into), and how you get rid of them, why they don't stay in there forever. I call it "the body's plumbing," which is a little bit crude, but then I'm a little crude, so that's fair. [Laughter.]

There are two general, big, polysyllabic terms that are used in the area and I've written them on the board: pharmacokinetics and pharmacodynamics. They're really "in" terms right now in the medical area. As far as their origin: "pharmaco-" means dealing with drugs; "-kinetics" is the motion; and "-dynamics" is the action and what it is doing. So pharmacokinetics is the term, the study, that covers the dosage that goes in, where it goes and how it gets there, what the concentrations are, how the concentrations build up, and how they drop off: It's the motion of the drug. Pharmacodynamics is the relationship between the concentration (the dose) and the effect the drug has.

There used to be a book on what's called medicinal chemistry, or pharmacology. Now there are books on the mathematics of pharmacokinetics, or the history of the study of pharmacodynamics. Big terms. Basically, this is all embraced in pharmacology. And basically, the whole course is, in essence, one of pharmacology's little smells of ethics and morality of opinion. But it's the talk of the department of chemistry, I'm not quite sure why.

I think one of the best ways of getting a view of how drugs get in and around the body is to get a view of the body, not as you know it with all the geometry, and arms and legs and odds and ends, but think of it as a bucket. How many people have taken embryology? Oh, neat! No one. Okay. [Laughter.] Now I can really get into a basic thing.

Embryology is one of the most exciting and boring things I've ever taken. Boring because it hasn't changed in years. Exciting because I had never seen it before. There is a phrase in biology that I had often heard but had never understood before, that ontogeny recapitulates phylogeny. Ontogeny ("onto-" refers to being, and "-geny", like genesis, refers to origin) means the origin of being, or existence of an individual. Phylogeny ("phylo-" refers to tribe, as in Phylum, a group that represents a series of separate individuals that represent a progression of evolution) is the origin of the individual in the sense of the family tree. The words are easily defined, but the concept is, each of us has two origins.

As the individual, we started a few years ago as an egg which became fertilized with a sperm and became a single living cell. When one cell becomes two cells, becomes four, becomes a bunch, and begins taking shape and you see the cells developing into an organism that gets lumpy, and then the thing evolves, and the nerve crests close over. Pretty soon it looks like a bitty thing, then it looks like a bitty higher animal, then it looks like a bitty human.

It's amazing how this process of embryology reconstitutes much of the evolutionary development of the human being. And you see it developing from its very origins in a one-celled body, the egg, to the two-celled body, the morula, all these different terms. You can observe the progression from the ancient and the simple, up through the complex, to the chicken, to the pig, to the human. Embryology is the few-week or few-month moving picture that relives (recapitulates) that entire history of the development of the organism.

The easiest way of viewing the body as a vehicle for the absorption and transportation of drugs, is as a sort of torus or donut shape. So I want to describe the body in the concept of a torus, where you have a body that's got a hole through the middle of it. Or it may be better to think of it as an apple in which you've the cut the core out of the apple. So you have the substance of a body, and a hole that goes completely through it. That's more or less what the handout for today is, with the funny arrows on it.

In the embryological development of a mammal, the growth can be seen to consist initially of only outside cells, and those with division and acceptance of individual character for this torus structure, that represents the outside surface. The feeling you have to get is that the outside of the apple, and outside of the lining of the hole through the apple, are all outside of the body. The body is a pile of tissue in a bag of skin with a hole through the center, a tube. The hole starts with the mouth; let's follow the tract. You go through the mouth, down the tube into the tummy, the stomach goes through a pylorus, it gets into the small intestines, or small bowel, this snakes around in all kinds of weird ways, and it goes up to a big bowel, the ascending lateral, the descending big intestines, it goes into a colon, a rectum, and an anus. So you have the top of the core of the apple that is the mouth, and the bottom of the core that is the anus. Anything outside the skin is not inside the body, and anything that's in this tube is also not inside the body. You can't see it, it's dark in there. But it's not inside the body, except in the sense that you can kill the bugs that live inside the gut.

Generally, if you were to swallow an olive pit, depending on your particular mechanisms of peristalsis, in a few hours or a day or so, you would find deposited in the toilet an olive pit. That olive pit has gone through the body, but it's not done anything to the body. It's never gotten inside, in the sense of pharmacology. A marble chip will go right through, so that is outside the body too. Any drug you put in this way, and it comes out that way, will not have an effect because it's never been inside the body. To get inside the body you have to get through this outside surface, or through the inside surface, in some way. The mouth is a common way of taking a drug, but unless the drug gets absorbed in the body, it never has an effect. I'm talking in generalities. There might be something happening within the gut that affects the body, such as killing bacteria, which then release some toxin that is absorbed, but in general this is true. Everything I say is 10 percent false because there are exceptions. But in general, get the feel now of the music and then get into the exceptions with a little bit more experience and exposure later.

Let's look at a tissue that is not part of this hole in the center of the body. You've got a heart, and it pumps blood all over the place. The blood is the transportation system of food, of oxygen, and in the interests of this particular course, of drugs. How much blood is there in the body? The blood, after all, is the tissue that almost always is responsible, one way or the other, for getting a drug from where it's going in to where it does something. You inject a drug into a vein, you take a drug in the mouth, you give a suppository in the behind. The receptor site is not in the arm, or in the mouth, or in the rectum. As a rule, the drug must get into the blood somehow to get to the location where it will be active. The receptor site is in the head, or where the offending bacteria is you want to kill, or where the emotions that you wish to change are, but this target is rarely in the blood itself.

The drug has to get from that insertion into the body, transported around, and deposited where it needs to go. The drug does not know where it wants to go. Do you have drugs that seek out a specific receptor and somehow by magic accumulate there? Generally, no. Generally, drugs go everywhere through the body, they follow a distribution, like dye in a cloth, staining everything, and they get distributed everywhere into the tissues by means of the blood. They come out of the tissues, and the body has a way of getting rid of them. The body gets rid of everything. It doesn't know a drug from a food. The body takes in and gets rid of. And a little bit, only a fraction of a percent of the drug, gets to someplace that rings a bell that achieves a drug action. And so, often 99 percent of the drug is absolutely wasted because you have to put in a hundred times of it to get one part, 1 percent of it, to where it's going to be effective.

There are exceptions. There are drugs that are seeking out a specific thing, and that specific thing is something that you want to change, and you can use these specific drugs. Iodine will seek out the thyroid because the thyroid has a great big sponge that says, "I love iodine." If you put radioactive iodine into the body as the iodide ion, you can cook the thyroid. That's one of the methods of thyroidectomy. You have a person whose thyroid is out of control, cancer of the thyroid, you give them a great big charge of radioactive iodine. It doesn't cook the whole body because it's all sucked into the thyroid and cooks the thyroid, completely destroying it. So there are drugs that are target specific.

But most drugs are strictly in there. You want to water the beans in the garden, you water the whole territory, and enough water gets on the beans to satisfy and the rest of the water is thrown away. Generally, you have this throwing of the drug into the body and that little bit that gets to where it's active is what counts. The rest of it you say, "I rather hope it doesn't affect anything else negatively."

How does this work? You have a heart, you have blood that courses through the whole body. How much blood do you have? Well, if you exsanguinate it, that means you take out all the blood you can; you put in a needle and keep drawing, as long as the heart pumps out blood, you haul it out, and you would have a dead patient on your hands. If you exsanguinate a person, you'll take out about five or six liters. Let a liter be perhaps something around a quart, that's maybe about a gallon and a half of blood. That's the inventory of blood in the body. The heart does its pumping job, and what it does is it pumps blood out, under pressure, into tubes that are known as arteries that go through capillaries out into every tissue, in the liver and the spleen and the brain, wherever. They go into capillaries, which are continuous to capillaries that now drain the blood back. This is known as the venous system, the veins that collects the blood and shoves it back to the heart. The heart pumps it around again. It's like an air conditioning system using the air over and over again: using blood over and over again and pumping it out.

So, you have about five or six liters of blood, and in general the blood makes the whole tour around the body. You can't say this is exact because some paths are shorter than other paths. It makes the whole turn around the body in about a minute. So in essence, if you were to inject a drug, pull your needle out and put in a probe that could detect the drug, that drug you injected would disappear and you would see the first signs of it coming around again in about a minute. So you have a problem if you want to get blood with a constant concentration and you want to do it by injection. You might inject either very slowly, so you ooze the stuff in over a period of time (the steady addition of a drug at a slow rate over a long period of time is known as an infusion) in order to build up to a constant level, or you put it in as a bolus (a bolus is a compact volume given in a short period of time that is experienced all of a sudden) and it becomes more diffuse as it goes around the body. Some gets back to the heart earlier, some later. And pretty soon, the first is showing in about a minute, and it takes about two or three passes around the body to smooth out the concentration. So you're going to get surges and all kinds of uneven concentrations for a minute or two or three. And then it finally smooths out. So the heart, by definition, pumps about five or six liters of blood a minute. That's the pumping rate at the volume of blood, and the blood makes a tour in a minute. You may think of blood as a liquid, but blood is a tissue, just as surely as a muscle or bone.

Let's take the most basic approach, most people take drugs, use drugs, get their contact with drugs, by mouth. How does a drug that may be active in the brain get from the mouth down through this tube? Can you absorb from the mouth? Sure, there's a lot of absorption from the mouth. The mouth has a lot of fine tissue. Sublingually (placing a drug under the tongue) is a very common way to get things into the blood rather directly by not going through the gut. The gut, in its Latin term, its medical term, is *enteron*. If you insert a drug into a body in a way to bypass the gut, it's not as hard on it. As enteral administration, you can go in by mouth, you're *in entera. Per os* means by mouth. This is a normal, common, self-administrative, ethical (in our society) means of administering a drug. If you avoid the mouth, you avoid the gut. If you go in any other way, you are parenterally injecting the drug or administering the drug.

Let's look at some of the parenterals before we go into the mechanics of enteral ("par-" or "para-" meaning across from, beyond, or aside, and "enteron" meaning the intestines.) Under the tongue. If you look under the tongue, by god, it is a big, bulgy thing with two heavy, big blue veins and all kinds of blood right near the surface. And substances are absorbed through that tissue directly into the blood. The reason you want to get in sometimes without going through the gut is it's fast. You need a drug that immediately does something. You have a person who's in a bad situation and you have a counter-bad of some kind, you want to get the drug in, and in quickly. Self-administration by under the tongue is a relatively fast way. Of course, a certain amount of anything put there will trickle down the throat and become an oral administration.

STUDENT: Is there anything quicker than IV?

SASHA: No. No, it's not as quick. The interarterial is the fastest. IV is quite fast. It is not as fast if you have more tissue to go through to get to the blood. The more directly you're in the blood, which is intravenous, where you go directly, and usually go in a vein, wherever you can find a vein, you go in and you're immediately depositing the material in the blood. That is probably the fastest method unless you want to punch a hole in an artery, which is always a risky thing. Arterial injection is used for true emergencies only. But in going through a vein, you can get various veins draining out of the head, you can get veins in the arms that are easily available, veins in the hands are easily available, veins in the groin are easily available. And going in that way, you are depositing the drug directly in the vein and, as such, it is already within this one-minute tour where it's going to go to where it's going to have its action. This is the advantage of speed.

The big disadvantage is that once a drug is deposited in the blood, or muscle, or spinal column, it cannot be recalled. It is like a submarine-launched missile, irrevocably on its way. Once it's in the vein and been put in there, you can't suddenly go upstream a ways or downstream a ways and put up a blockade and say, "I made a mistake. I miscalculated by a factor of ten. Give me my drug back." Can't do it. You can orally. You can make people vomit, you can pump the stomach. Very few drugs absorb from the stomach. They have to go on through the stomach and into the small bowel. And there is where most of the absorption occurs.

Let's go on with the parenteral administrations. Under the tongue, up the nose. What's the technical term for taking a drug up the nose? Cocaine snorting. Do you know? Insufflation. It's quite a neat term. "Snorting" is far more common. [Laughter from class.] You have soft tissues up in this area of the nares, and absorption again into the blood by going through a very fine tissue. A lot of that which is taken up the nose drains into the mouth, into the stomach, and becomes oral.

[Directed to student] Yes.

STUDENT: Can you spell that?

SASHA: Insufflation. Good God!

ANN: I-n-s-u-f-f-l-a-t-i-o-n. I think.

SASHA: Sounds reasonable. Snorted. [Laughter.] How else can you get in? You can go through any hole in the body. You can go through any tissue, inject any tissue. Put something on your hand. If there is an intermediate that will carry it through the skin you have a percutaneous administration through the skin. You inject it under the skin, skin popping, is subcutaneous absorption. Stick it into a muscle: intramuscular. Here are two terms: "inter-" and "intra-." These prefixes are often interchanged but they are absolutely opposites. "Inter" means between. Obviously, you do not make an "intervenous" injection unless you're aiming for a vein and miss it, so that you go between veins. You make an "intravenous" injection because you want to go into a vein.

The way I always remember these two terms is "interstate commerce." I've never heard of "intrastate commerce." It's "interstate commerce." That means commerce between states. "Intra" means within the state. So "inter," between; "intra," within. So if you're injecting into a vein, you are giving an intravenous injection. And, of course, that is a way you get directly into the blood: Put it in the blood. If you want to go in the arterial direction, you make an intra-arterial injection. Primarily because arteries carrying blood are highly pressured, are much thicker, much harder, and more hazardous to enter. In about one time in ten, or one time in twenty, when you put a needle through an artery and inject into an artery, the person who owns the artery will go into shock. It's just a response of a very deep trauma to the body. And hence, it's only used in emergency situations where it's absolutely necessary and you have facilities for handling the hazard of shock.

STUDENT: What about suppositories?

SASHA: Suppositories. That is kind of a twilight zone. When you administer a drug rectally, through the anus into the rectum, it's called a suppository. Does it or does it not serve a parenteral function? Technically, it is part of the gut, so, technically, it is not parenteral. But the parenteral thing is primarily to avoid the gut absorption which goes to the liver. And so it largely depends how it is absorbed from the rectum. If it is high enough, it will be caught up in the splanchnic circulation and will go to the liver. If it is low enough, it will not. It will be caught up in the renal return and will bypass the liver. So it's a mixed message. It depends on the nature of the drug and the depth and the nature of your anatomy. It's the depth of the deposition of the suppository that will dictate whether it acts as if it were parenteral or not.

Other routes. Drugs can be given through the lungs by inhalation or by smoking. Intramuscular is injection of the drug directly into a muscle. Muscle is very slow because the drug gets in the muscle and it has to be picked up by the capillary bed and be filtered back into the venous system. Another parenteral route is intraperitoneal where the drug is placed on the delicate tissues of the mesentery net that surrounds the intestines. And with a long and flexible enough needle or tube (catheter), the drug can be directly placed in extremely remote and hard-to-see locations. There are other parenteral routes I have not listed here.

[Directed to student] Yes.

STUDENT: Do they ever inject right into the skull?

SASHA: Through the skull? There's trepanning, which is drilling a hole in the skull as a way of getting in there. Remember, the tissue within the skull, the brain tissue, has almost no nerve sensitivity to pain. And almost all surgery that involves cutting a hole in the skull, pulling out part of the bone and digging in there and doing whatever you're doing, removing tumors or such, is often done with a local anesthesia. Because there's no pain inside the brain. Evolutionarily, there's no requirement for building up sensitivity inside the brain because there's no normal circumstance of getting there.

STUDENT: But no action in terms of dealing with some sort of nerve response in the brain?

SASHA: No. You have to use a different distribution for getting to the brain. Arterial, using an intra-arterial injection would be your best bet, into the arteries leading to the brain if you want to feed the whole brain. You can give an intraspinal injection, which is close, and go between the bones and enter the spinal column itself and get directly into the cerebrospinal fluid. Very fast acting.

[Directed to student] Yes.

STUDENT: But isn't that very dangerous?

SASHA: Absolutely. Once you begin shoving needles in and around neurons, you have to know what you're doing. Remember that spinal anesthesia is often not entering the spine, but injecting anesthetics around the neuron synapses that are leading out of the spine. When you enter the spine, for something like a spinal tap where you want to determine, for example, evidence of what's called a CVA, a cerebral vascular accident, where something pops a vein in the brain and you actually get blood into the spinal fluid, into the fluid of the ventricles, it can very rarely be seen by looking at a spinal tap. There are no cells of any kind in spinal fluid normally. The thought that is usually used: the presence of any cell is pathological. And so if you find brown sludge, or cells, or aspects of particulate tissue in spinal taps, you know that there's been a ruptured vein or artery in the brain. And that's one way you can follow the sequence of a stroke, a cerebral hemorrhage, to find out the cell inventory of the spinal fluid there, you actually enter and withdraw a sample.

Sometimes you enter and inject air, which is tricky yet somewhat valuable in certain circumstances. Twenty or thirty milliliters of air that goes *blllpp*, up the head and enters the ventricles, then you lean your head to one side and *bllluup*, the air goes into one of the ventricles, and then you take an X-ray. Since blood is a tissue and brain is a tissue, brain X-rays have never been particularly effective. You cannot see one kind of opaqueness against another kind of opaqueness with very much res- olution. But if you remove the fluid by putting air there, then you have something much less opaque. And so you can actually get a picture of the geometry of the ventricle by putting the head this way and taking an X-ray. Then you put the head the other way, *bllluup*, and take another x-ray of the other ventricle. Then you look at it and here's a perfectly fine ventricle, and here's a ventricle with a big thing pushed into the middle of it. You've got your clue that maybe that's where something's growing and that's where something's going on. The person will eliminate the air in about forty, forty-four, or forty-eight hours, but will get a blinding headache! It is a rebellion against that procedure that you get a nasty headache for a day. But you can make parts of the brain visible by X-ray, which you cannot in any other way.

You can make it visible now by methods known as positron emission tomography, PET scanning. And when I get into the lecture on radio- activity, I'd like to talk about positron emission tomography and NMR scanning and CAT scanning, these methods of getting a view of what is going on, how you can see things in the body and how they work, and what the good aspects are, and what the limitations are.

Everything that's new has no limitations for the first five years. Any drug that's brought onto the market for the first five years is totally effec- tive without side reactions and without toxic problems. At some point between the fifth and tenth year, they start to be less effective than you thought. You find they have side reactions you were not aware of, and find that toxicological problems are just coming to light. But fortunately, there's a new drug that comes out about that point and for five years it has no side effects and is totally effective in every way and has no toxic issues. So I've heard it said, in one lecture in medical school, use a drug in the first five years of its introduction because that's the time it is totally effective and has no side effects. [Laughter.] It's there, in the first promotion of a machine, of a device, of a drug, it's a panacea, it's the thing that's going to bring the answers. It doesn't turn out to be that way, but it's a good start. While people believe that this is a cure-all, it will have a lot better effectiveness because there is a placebo effect that is very real. Say this is a new thing that's proven to be very effective in the treatment of whatever you're being indisposed by, it has a better chance of being effective. You've got confidence, you've got it from an authority wearing a white coat and stethoscope, and that is the authority figure who says, "This drug is a new one and it's very effective." And it has a good chance of being effective.

[Directed to student] Yeah.

Student: Are we actually learning by drugs not being 100 percent effective not to trust the medical staff?

Sasha: No, I think the placebo effect is stable. It's very effective.

[In response to a student interjecting that the effectiveness of placebo was based on the belief it is effective] That is why the term placebo is not used. In fact, one of the most pejorative adjectives that I've been fighting against for years in the drug area is the pejorative adjective "just." "It was just a placebo effect." Tearing down the whole concept! My golly! Darvon! How many people have taken Darvon? Okay. One, two, three. Has it been effective? Nah! You look at the curvature of Darvon, it's a wipeout. It's not a particularly effective anesthetic, but it was a superb placebo reactor for years. And you took it with the confidence it was going to do the job. When you put it in an animal or put it in humans in a blind study, it's not particularly effective. It's a nice example of sales hype, and of confidence in your physician who's acting in good faith!

Because they read the last five years of literature, and they know that it's an effective anesthetic, an anesthetic and analgesic.

STUDENT: It would also depend on the concentration.

SASHA: Depends on concentration and what the problems are that you're facing. Concentration. In fact, that's what I'm getting at in the whole thing, is how drugs get in and where they build up.

We've covered a lot of the parenteral routes. Let's follow the enteron. If you were to take a pill by mouth and swallow it, it first resides in the tummy. Some things in the stomach are absorbed in the stomach. Alcohol is a good example of an easily absorbed drug, for if you were to rapidly swallow two or three stiff drinks you would feel the effects of the alcohol resulting from direct absorption within a very few minutes. So there can actually be some absorption in the stomach. As a more general rule, passage from the stomach on through the small bowel (small intestines) is needed before absorption occurs for most drugs. Remember, the inventory in the stomach is first held there, and it takes a while for it to vacate through what's called the pylorus, a little valve at the bottom of the stomach that often works and sometimes doesn't work. Four hours, eight hours after a meal you *urrp!* a little bit and you can taste the ham that went down eight hours earlier. Clearly, the ham has not cleared the stomach if you can *urrp* it up from the stomach. The quantity and the nature of the stomach contents can severely affect the rate of absorption of a drug taken orally. As does a person's state of health, as this can also be reflected in the residence time of food in the tummy. Usually in half an hour, it drains into the small bowel from where the absorption takes place. This small bowel, the small intestines, is this convoluted thing with peristalsis, pumping things through it all the time. Fairly basic. The stomach is quite acidic. Most drugs are basic and will not be absorbed through the tissue under acidic conditions. They're ionized, they're charged. But once they're in the small intestines, under a basic environment, they're not ionized, and they'll be absorbed readily and efficiently though the walls and picked up directly by a capillary network in the peritoneum.

From here, the drug is picked up and transported by what's called the hepatic portal system. This is really the heart and the central mechanism of drug absorption. Here's a hepatic view of the liver [referring to drawing on the board]. Portal means moving, transporting things. There are about three portal systems in the body. The major one is the hepatic portal. The term portal is an unusual thing.

Arteries are defined as the tubes that the heart is pumping blood out into under high pressure from the heart, going out into capillaries somewhere. Veins are the less strongly built tubes that go from capillaries back into a big tube that eventually returns to the heart. There is a pump in an artery. There is no pump in a vein. Veins are transporting the blood under very low pressure, often in the *inferior vena cava*, almost no pressure, back to the heart. Veins do not pump. The body depends on muscular action and tissue flexing to move the blood back to the heart. That's why when people stand at attention protecting Buckingham Palace for four hours on a warm day, they often fall over in an unconscious faint, because the blood accumulates down here [indicating the feet] and there's no pump to pump it up to the heart. When you're walking, you're flexing muscles, you're moving, all this sort of thing, there is a general moving of blood back up. I mean, water doesn't flow uphill. And so if you're standing absolutely still and your feet are down and there's no muscular action, you are intentionally immobile, the blood will accumulate down here and you run out of blood. There's no blood up here. There's no blood up here, none gets to the brain, over you go. It is nature's way of getting blood back into the head. It uses gravity, you're lying flat, there's no uphill/downhill anymore. Now you're flat [laughter from class] and the blood gets back to the heart and goes up to the brain. That's why when a person faints, you get their head down. The fainting is because there's no blood in the brain, not getting air, not getting sugar. And so you put the head down, water flows downhill, and very soon blood will get to the brain, even with insufficient heart action. So you find a person going over in a faint, don't be hesitant to put their feet up a little bit and get the head down. It's much more responsive with time than just letting them sit in a chair.

So, you have arteries going out under pressure, veins flowing back in from capillary to heart. There is a third type of tube, often called a vein because it's not an artery, and its common name is a portal system. A portal system is the structure that starts from one capillary bed and ends in another capillary bed. It starts and ends as capillaries and there is no pump in the system. The heart is not attached at all. And you have this type of system in the hepatic portal. The capillary bed at the coming-in end is fed through capillaries from the arterial system that feeds the intestines, the mesentery blood flow, which is the gathering of things out of the intestines, and gathers in this capillary net coming into the hepatic portal. You also have this renal branch of the hepatic portal, so the whole thing is called the splanchnic system. All blood going to the liver is called splanchnic circulation, not necessary for you to know. But the main part of it is in the hepatic portal. Some of it is in the splenic, which is from the spleen, which is a good inventory of blood into this portal. And, of course, you have arteries that feed into the liver flow. But the main point is you have this tube that goes from the intestines where all the nutrients are, sugars and drugs that have been taken orally, gathered in this portal system and delivered to the liver where it diffuses back into a capillary bed.

What is hard to believe is that this portal system consumes about a third of the energy of the heart. Of the five or six liters of blood that's pounding around the body, about a liter and a half of it every minute flows through this portal system. That is a major transportation. That's how all your nutrients, all your food, all your digested goodies that get into the small intestines, get into the body. They are absorbed through the intestines into this portal system, gathered up into this portal vein, and are delivered to the liver where they are then worked on. The liver is a tremendous factory for chemical change. And many drugs are quite distinctly chewed up in the liver. If you take a drug into the liver, maybe only 30 percent of it gets out.

So this is why often you bypass the guts if you want a drug to have maximum activity, because you want to avoid the liver, at least in the first pass. This has the two advantages of speed and intactness of administration. Anything absorbing from the intestines must pass through the liver, and this organ is a major metabolic machine. If you want a drug to get to the brain and you go through the mouth, It goes from the mouth to the small intestines, and from there it goes through this portal over to the liver, which says, "Hah! I'll take my 90 percent of the drug because this is an alien thing. I'm going to chew it up!" And it chews it up. Ten percent escapes and a certain ratio makes it to the brain. If you want to get a drug to the brain quickly and with higher efficiency, you'll bypass the liver. You'll bypass the liver on the first pass, it'll go to the brain. How long does it take to go from, let's say, this vein to the brain? Fifteen seconds.

Let's follow the course of the drug from the heart to the brain, whether it gets to the heart directly by intravenous injection, or indirectly from the mouth to stomach to intestine to liver via the above-discussed scenic tour. Blood containing the drug is carried back in, it goes to the heart. The heart in most higher animals (mammals, birds, some reptiles) has four chambers. Take your right hand with the fingers pointing down over where the heart is, about the middle of the chest, and touch near the knuckle of your fifth finger. That is the right atrium. That's where blood gets back to. That's where the venous blood feeds back to. It gets pumped down to the right ventricle, that's your little fingernail. It gets pumped from there to the lung, it goes directly from there through a system to the lung. And remember it's being pumped, so that is an artery. Some people say arteries carry oxygenated blood, veins carry deoxygenated blood. Nonsense! Arteries carry blood away from the heart, oxygenated or not, and veins carry it back to the heart, oxygenated or not. The blood going from the lower part of the right-hand side of the heart to the lung has no oxygen. That's why it's being sent over to the lung. It needs oxygen again. But, it is an artery that's pumping it, the pulmonary artery going over to the lung. The lung it goes into is a capillary network that gathers up air, dumps carbon dioxide, and gathers up oxygen that diffuses into the erythrocytes (red blood cells). Then it comes back from the lung in the pulmonary vein, into the upper left part of the heart (this is where your thumb knuckle is located). This is called the left atrium. At which point, the blood is pumped down into the left ventricle and then out into an artery. And it goes out that artery. The three main calls of that blood: the brain, the heart itself (the coronary arteries are one of the major unobstructed artery calls of the heart) and gut system. That's the handout I have here. The arteries will feed all tissue. Look at the handout for today. It's diagrammatic. The heart is not really shaped like a Valentine

The vertical tube is the core though the apple, the alimentary canal, the gut, the hole through the human body. And the flare of fine lines feeding into and away from it, with the gut and all the other organs, represents the capillary network. You must remember that this system is not discontinuous (it does not go in and end, then start up again to go out) but rather it goes through the organ. Blood perfuses each organ, but it is always contained inside of a tubular structure. There is no free blood in any organ outside of the heart. When an organ or tissue is injured and blood flows from it, this is due to the actual rupture of innumerable

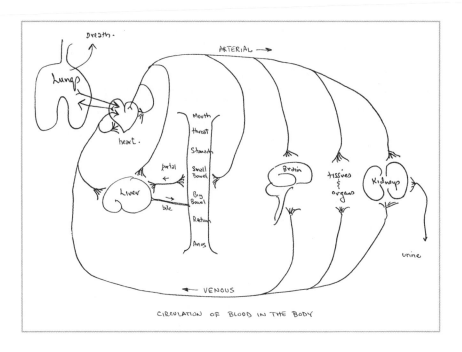

capillaries (or perhaps larger structures such as veins and arteries) and these are the sources of the blood.

This is the sort of lecture I love, where I sort of ramble about things. The notes I have are sort of minor.

What is the shortest time it could take for a drug to be active? There may be delaying factors such as the need of prolonged exposure at the target site, and/or biotransformation from one form to another in the liver or some other metabolic tissue. But the response time cannot be any less than the time it takes to get from the site of administration to the target organ itself. How long is this?

If you take a compound orally, how long will it take to become active? There may be some absorption from the mouth, and from the stomach, but in general there is the requirement to get to the small intestine. Within the stomach, if you've just gone through a milkshake, cheeseburger, and some french fries, it's apt to be a bit behind the crowd going into the small intestines. And if it's not absorbed from the stomach, it may take a half an hour to an hour, or even longer, for it to wander into the small intestines where absorption can occur. From there, things move pretty fast. If the drug is placed directly into a vein, the elapsed time to reach the brain (if that is the site of activity) may be only a fraction of a minute. And with pulmonary exposure (from inhalation) only a few seconds are needed. And if you enter the blood supply feeding the target organ (as with an arterial injection) the availability is immediate.

Once a drug is absorbed, it's going to be transported like a shot over to the liver, but you are still subjecting the action to the transformation of the liver. Now don't get the idea that a drug is a drug and once the liver gets hold of it and changes it, it is no longer a drug. It can be quite the opposite way. A material may *not* be a drug and may depend upon the chemical capability of the liver to change it into something that *is* a biologically active drug. So you cannot say parenteral is always faster or more effective or more active than oral. Oral can be more active than parenteral if you need to have the liver's good offices to make the drug an active drug.

So, there are materials that are called prodrugs. Why "pro?" I don't know. I think they should be called "predrugs" because they are not active, but get converted into things that are active. But in someone's infinite wisdom, they are now called prodrugs, where they do not have what is called intrinsic activity, a term you'll hear that is a little tricky in its use. Intrinsic activity means that the drug is active and has the capability of producing some response as being one of the properties it possesses. If you knew where a receptor site was, and could get to it, then took a saltshaker and sprinkled the drug on that receptor site, the receptor site would say, "Ha ha! I'm going to do something." That is intrinsic activity. No change (whether from metabolism in the liver, or the lung, or the blood) would be needed.

Often a drug has to go through what used to be called detoxification. This change of a drug that the body is capable of making has been called by many names. These are valuable terms because of their misleading indications. There are two reasons the term detoxification has become very unpopular. Detoxification, we speak of the liver as being the big detoxifying agent, implies something that's toxic will go into the liver and get converted into something that's not toxic. That's utter nonsense! Sometimes things are not toxic and get to the liver which makes them into things that are toxic. Sometimes toxic compounds are turned into things that are more toxic than the original compound. The liver is not a great big super-intelligence with a great brain attached that says, "Aha! That's a toxic thing. We'll do something to it," or "That's not a toxic thing." The body does not know how toxic something is, and somehow knows to change it to something less threatening. And secondly, the body does not know that a given chemical is even a drug. It acts in its generalized way on everything that comes to it, be it drug or food. So the thought of making something less toxic, that is implied by the word detoxification, is completely misleading.

A replacement word is metabolism. One speaks of metabolic capability, metabolic pathways, and rates of metabolism. But again, there is the implied suggestion that you are descending from something important to something that is simply being readied for excretion.

The term now used is biotransformation, which is explicitly accurate. Changes that occur in the body. There is an important concept that should be understood. This applies to targets of action, whether one is speaking of the action of this new drug on some receptor site, or the biotransformational assault of the liver on a drug. The body, the liver, the sites in the brain, have no advance knowledge that some big manufacturing company is about to market some Thioburpalene that will lower the blood pressure. There may be a site of action, but it has not been created during genesis simply for the anticipated coming of Thioburpalene. And the liver may be able to synthesize the more easily excreted trans-3-hydroxy-thioburpalene. The drug company Smith, Kline & French has just turned out a new SFK23394 that's very neat, it's a possible beta blocker and may have action for the heart. You put it in the body, the body doesn't say, "Oh! I'll send that to my SKF23934 receptor." [Laughter.] People speak of PCP receptors in the brain or LSD receptors in the brain. God, the brain was around way before you ever had PCP and LSD. And the body has not developed this magic in the last eighteen years since these drugs were discovered. The body has been around since long before these drugs ever existed, and the action and the disposition of such drugs simply follow their presentation to the normal bodily tissue.

The body has the capacity to, and will change if necessary, everything it gets, be it drug or food. Why does it metabolize it? Basically for one reason: All the mechanisms of the liver's action are geared toward making a molecule, a drug, a material, more soluble. The whole direction of the liver's change is to solubilize something. A common trick is to add a sugar like glucuronic acid, which really increases water solubility. Why? Because the major sites within the body are associated with tissue, and there are membranes to be gone through, and fatty barriers to be reckoned with, but most stuff we take in gets peed out through the kidneys, and that's a completely aqueous system. The ease of getting rid of something is increased in direct proportion to its ease of solubility in water.

There is a term in the chemical and biological world that describes the property or moving towards the fat or moving towards the water side of a two-phase system—it is partitioning. This is a measure of the relative concentrations that a compound or substance will establish when distributed between two insoluble things, such as water and oil. The oil-ness is more or less needed to get into the body and show activity, and the water-ness is needed to be gotten rid of. So the biotransformational procedures of the body are to add oxygen, usually in the form of oxida-tion (hydroxylation, making sulfoxides, carboxylic acids) or increasing ionizability (as in making strong acids that can form ionic salts). We also mentioned adding a sugar like glucuronic acid which really increases water solubility. In this way, the compound (drug, food, chemical) can be more easily cleared into the urine and excreted.

The main avenues for excretion are the urine, by definition containing water-soluble things, and the stools, containing materials never absorbed by the body plus what is contributed by the liver by way of the bile tract. Minor routes include the lungs, removing things from the blood that are volatile, and perspiration, removing the things that are secreted into sweat and passed through the skin.

A dramatic example of excretion through the skin can be seen with the historic treatment for worms, elemental sulfur. One such disease is schistosomiasis (bilharziasis) which is caused by some species of trematodes (aka blood flukes). Schistosomiasis will come into an area from where it is endemic; Asia and parts of Africa. For example, it is endemic throughout all of the rice-growing cultures in the southern part of eastern Asia. If you could only somehow get the people to not wander around in the water in the middle of the day, when the schis-tosomes are active in the snails, and are actively in this vector cycle. But that is the kind of thing that requires convincing, education, and disruption of the economy. People work in the fields all day long; they are not about to knock off for three hours in the middle of the day, because there is no concept of what the Schistosomiasis is. It's a worm that is in there only because it is in residence for food. For your food. When you eat, one, two, three, there goes the worm. And it is a very common disease. It is one of the major worm diseases of the world.

And again, it requires this vector cycle of getting into the snails and carried into humans.

The sulfur loading of a person or an animal with intestinal worms is an ancient treatment used for treating the worms. When you take a large volume of sulfur it is a rather amazing thing. You can take a person after about a day of sulfur treatment, and if it is a dry day you can roll up the sleeve and run your hand down their arm to release yellow clouds [of sulfur dust]. The sulfur is in there to such an extent it is actually excreted through the pores and it ends up on the skin.

Another form of excretion is through the breath. If something is small and volatile and it's carried in the blood, the lung is a superb gas exchanger. And if it's something that is gaseous or has a high vapor pressure, it can be excreted through the lungs. In pharmacological studies there is often the need of administering some insoluble drug intravenously, and suspensions can be made in some form of wetting agent. But the injection of anything that is particulate is very hazardous, since all blood must eventually pass through a capillary system, and the insoluble will be filtered out and thus clog a capillary. This is one of the problems associated with the use of insolubles such as talc for the cutting of illegal drugs, or as an ingredient of pills that are intended for oral use but are instead injected, the talc is hard to remove from the injection bolus, and once in the blood stream, cannot ever be removed from the body. Each particle will clog something, and the most noticeable damage from clogged capillaries is in the retina. Here, with a capillary not being able to deliver oxygenated blood, a group of cells will be damaged, and there is a small but measurable loss of sight.

A lot of solids can be taken in one form or another. One of the rather interesting things that I had heard was being used during the late 1960s in Haight-Ashbury was injecting insoluble drugs dissolved in methylene chloride. It will dissolve things that are lipophilic and won't go into water. They would dissolve something like N-N,Dimethyltryptamine (DMT) in methylene chloride and inject it into a vein. So you suddenly have this bolus of solid—going through the vein, into the heart, turning around—that is not totally insoluble but is largely insoluble in water. And once it's in there and it gets to the lung, it goes out through the lung, and you exhale this cloud of methylene chloride. The drug doesn't go through the lung, it doesn't get exhaled, so the drug is left in the body. It's a way of getting an insoluble drug into the body and it's actually been used now clinically at Lawrence Lab.

[Largely inaudible student question concerning the toxicity of chlorinated hydrocarbons]: Methylene chloride can be handled quite reasonably. Chloroform is more toxic. Methylene chloride is probably one of the safest of the chlorinated hydrocarbons. It is now being looked at as a possible carcinogenic agent. But it does not have the mischief of the other chlorinated hydrocarbon compounds. And you can get a material into the body that way, with the solvent released through the lungs.

STUDENT: But wouldn't the methylene chloride be fat soluble, in the fat?

SASHA: Mm-hmm. But you'd have this partitioning from the fat tissue. In fact, that is a major point of drug distribution. A very nice term is what's called "volume of distribution." It's a tricky one. When a drug is put in the blood, whether by an oral or by a parenteral route, one might think that its concentration is the simple division by the volume of the blood. Indeed, if all of the drug was restricted to and contained only within the blood, this would be so. But the drug is partitioned out of the blood into tissue, into fat, into areas that are in equilibrium with the blood, but are not blood itself. Thus, the concentration that is seen in blood will be lower, and can be very much lower than that which would be calculated. It's as if there were an immense pool of blood that had dissolved the drug, a pool even larger than the body itself. This hypothetical volume that can be calculated from the dose and blood concentration is called the "volume of distribution." It is a measure of the extractability of the drug.

In a sense, this is a prediction of the duration of deposition in the body. Let's say you were to inject 100 units of a drug into blood, and then you pulled out a sample of blood and saw how much material was there. If the material is so water soluble that it goes into blood, and stays in blood and doesn't go anywhere outside of blood and is forever in blood (one of the samples that's used in radioisotope study is rubidium, which does not emit too big a charge, the ionic thing won't go out of blood into the tissue), the volume of distribution is the blood. And so, you say, "This drug has a volume of distribution of five liters." If you have five liters of blood, that's where the drug is. And of course, if a drug is rapidly cleared, very rapidly by the kidneys, it has a short half-life in the body. On the other hand, a fat-soluble drug such as dichlorodiphenyltrichloroethane (DDT), is largely extracted from blood, and has a volume of distribution of perhaps thousands of liters. And one could predict that it would have a very slow turnover and long half-life. We will touch on these matters when we talk about THC, tetrahydrocannabinol, the major active component of marijuana.

A drug that has a large volume of distribution need not be sequestered in fat—it could be bound, for example, to proteins that are in blood. There are iodine-containing drugs that are opaque to X-rays and are extremely tightly bound to the proteins that are a normal component of the blood. The half-life of some of these drugs can be many months; they are being taken up by a new protein when the old one is destroyed by the body. They are not biologically active, since being so tightly bound, they cannot pass through membranes into tissues where an active site might be.

If you take something like THC, which has a high fat solubility, high tissue solubility, and take in 100 units of THC, we'll find only three units in the blood. So you can say 97 percent of it was in the blood and went somewhere else. So volume of distribution is the volume that it went into, had it all been blood. If it stayed in the blood, it'd be five liters. But rather than five, it's a distribution of 100 liters, 1,000 liters. It's as if the concentration you saw in the blood were a thousand liters of blood. There's only five liters of blood, so five per 1,000 is in blood, and the other 995 per 1,000 is out of the blood. And usually it's in the fat, in the tissue. It gets into a form that cannot be gotten from the blood. It can be bound. You have tissue binding in the blood because you have cells in the blood. There is protein in the blood, and you can have binding to protein. But if something is bound to protein in the blood, it's not available for biological action, unless the action is on that protein. But usually it's not, and so this kind of binding removes it from availability for action.

[Directed to student] Yes.

STUDENT: When you put 100 units in the blood and you get 100 units back out, is there a scale, some kind of a percentage? How do you express that?

SASHA: Using the volume of distribution. That's the point. Using a measure of how much of it stayed in the blood, and how much of it went via the blood to some depot or somewhere that is not blood.

STUDENT: Okay, so let's say you've put 100 units in the blood and you've only got three, what's the volume of distribution?

SASHA: One hundred threes of the volume of the blood. So it would be thirty times the volume of the blood, or 150 liters. And this is why you have problems with some drugs in which they are deposited in tissue and seep out slowly. I mentioned this earlier with DDT. There has not been the legal use of DDT in this country for some years. And yet you've got DDT in your body. You've got a pretty good inventory of it. And since it's in the body in tissue depot, usually in the fat of the behind, or fat in general, it's all the time getting into the blood and it's always in the blood level. The blood level is probably dropping, but dropping very slowly because the bulk of it is in that depot that is only in a very small dynamic equilibrium to the blood. DDT levels probably will drop over the course of many months or a few years.

One of the problems with the THC analysis is a lot of that goes into tissue, into fat, into an unavailable form, and you don't have turn-on pleasure receptors in the fat of your behind. So you're not, obviously, going in there to be active, but the body doesn't know what the drug is supposed to do, and it gets distributed in some way throughout the body. And a lot of it gets put in this neutral store, but it's all the time going back into blood. Your blood level goes down over a longish period of time. And therefore, after an exposure to something like marijuana, you can pick it up for days or even weeks afterwards because you have a very sensitive test, and you're picking up this distribution from fat or from another tissue, back into the blood.

This is a good point to bring up, a very critical aspect of body geometry, and that is what's called the half-life of a drug. Radioactivity is probably the best known and easiest to understand example of where this term will apply: half of everything decomposes in a period of time; half of what's left decomposes in that same period of time; half again will decompose in that same period of time. That period of time is the time necessary to reduce the substance to half of whatever it is.

How many people are familiar with the x-y axis? Fantastic. Let me portray this graphically. I am taking the moment to draw this out—I once gave an introductory lecture on radioactivity at a college I won't mention, and had a startling discovery. It was kind of a beginning class in the college. And the first thing I did was to talk about half-life, and did this kind of a thing [drawing simple x-y axis graph on board], and this is activity, and this is time, and that's the relationship. And I was told by a professor afterwards who sat in on the class, "You know, it's a good lecture. I learned a lot from it. But you didn't quite judge your audience. It is very possible for this captive audience that it's the first time they have seen that [an x-y axis]." Most people in explaining some relationship will start with the famous two lines, not realizing that this is not an intuitively grasped thing, but must be learned by use. In most cases, the y axis (the ordinate) represents the measure of something, the x axis (the abscissa) represents the passage of time, and thus the curve drawn in the indicated space shows how the measure of something changes with time.

There is an important concept dealing with the body's relationship to drugs known as a half-life: Half of everything you have is gone after a certain amount of time, and in that same period of time again half of

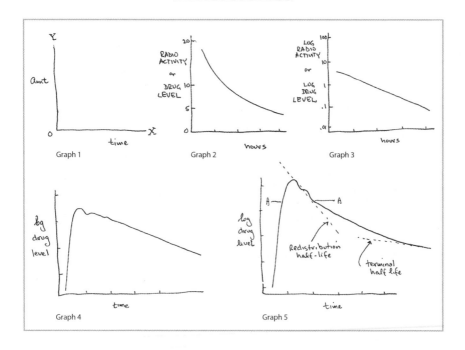

Graph 1

Graph 2

Graph 3

Graph 4

Graph 5

what remained is also now gone. The second graph shows the activity of an unstable isotope that has a half-life of one hour. Here's what it is at this time [referring to graph 2], when there is now only half as much left as was there at the beginning, and then here [referring to same graph] is where it is now, half of what it was at that time. There are 100 units of activity at zero time, fifty units left at one hour, twenty-five units at two hours, and so on. In both cases, this unit is called a half-life. Using a mathematical trick of taking this same information and making a log of the activity, you will get a straight line since it's going down a half unit of concentrated time. The slope of that line is the half-life [referring to graph 3].

By changing the identity of the y axis from radioactivity to plasma level, this graph would in principle show the blood level of a drug that had a plasma half-life of one hour. [In reference to graph 4] Drugs have half-lives in the body, but they are not as clean and neat as you would see in something with the concept of radioactivity. Because the half-life deals now, not with the drug going in, building up to a certain level and being excreted; instead, it goes into compartments that make it come out at different times. It may go into biotransformations. How do you look at a drug? How do you determine the log of concentration? By putting a radioactive marker on the drug? Then you're not following the drug, you're following the radioactive marker. And the radioactive marker then gets biotransformed into something else.

If a drug is metabolized, but the metabolite, the biotransformed form, still carries the radioactivity, you're going to be plotting, not the concentration of the drug, but the concentration of that radioactivity. And so this curve will reflect non-linearity because of biotransformation. And so usually, although this should be a straight line, and with radioactivity you define it that way, in the body it will have, usually, this kind of a curve.

[Directed to student] Yes.

Student: You defined law of concentration. But let's say, for instance, you do an intravenous bolus in, say, one milliliter. It depends on biotransformation, what metabolic process, because the concentration might stay the same with the drug.

Sasha: No, it will drop in time. Let's say you give an intravenous injection as a bolus of a drug, you're going to get this kind of a curve under log circumstances.

Student: So a peak and a—

Sasha: A peak and a drop off. And this is the final half-life. But this is not what's called a distribution phase. Because, once it gets in the blood, you're not immediately going to get the blood tissue back. It takes a while to do this sort of thing. And so, often there is more in the blood than you would think because it has not yet diffused out into the tissue. But once there's tissue out and there's tissue being back, the fat out and the fat being back, all are in static equilibrium, then you become linear with what's called the ultimate or the terminal half-life. You have a faster apparent half-life in here because this is not really the drug from the body. It's redistributing the drug into other parts of the body. You can take the drug in blood and the fact it has none, it's going to go that way [referencing graph 5] into fat. So it appears to drop off faster.

STUDENT: Yes, but if it's a prodrug, then you have to wait for the biotransformation.

SASHA: That's why I changed it to concentration of the drug. That's still, in the first part of this, dose concentration in terms of pharmacokinetics. We'll get into the other half as the second part of this. That first drop off was redistribution. Can you imagine the mathematics involved if you introduce a drug, go into a cell (a compartment it's called in the drug world), and from the compartment you have excretion? That would be straightforward. You have a straightforward half-life. But then you have this compartment that is in equilibrium with this compartment: K1: K2. And this compartment, by the way, is in equilibrium with two other compartments. Each of those is in equilibrium with something else. And so to know what's going in here, to calculate how much goes out the bottom requires you know all the dynamics of everything that's going on with the insides of all these other compartments. So you have this multi-compartment system. And sometimes, if you're following something going in by something coming out, and you are looking at something else that's coming out than what went in, then you have the mathematics involving biotransformation as well. And that's why that curve should be a flat curve, and usually it's not. Because when you get way out yonder when everything is in equilibrium, you get the other two states coming out. Then you have a flat half-life.

So, the half-life of a drug could be very long. I think I've mentioned one iodinating system for visualizing the heart where the half-life of the drug is substantially a year to two years. Because it goes in, and one of the fastest things it binds onto is the protein of the blood, and it doesn't let go. I mean 99.997 percent binding to that protein, but it doesn't have any biological action, because the protein will keep it from being active. But you're using the drug, not for its biological action, but for visualizing because it's opaque to X-rays and has nine iodines on the molecule. And of course, the protein goes where the blood goes. That's all you want. You want to see where the blood goes. So it's perfectly good as a visualizing agent, but it's in the body forever. Binding to the protein, gets metabolized, the thing is free for a moment, it gloms onto a new protein. And since you're all the time generating protein, it will stay on there, and it will have a half-life of a couple of years.

On the other hand, a very common drug used quite broadly in our society is heroin. The half-life of heroin in the blood is three minutes, it is metabolized and changed in the blood. It doesn't have to make it to the liver. The blood is filled with what are called esterases. The term "-ase" is a suffix to an enzyme that does something. An esterase is an enzyme that does something to esters. And the only thing you can rationally do to an ester is hydrolyze it. And heroin is a di-ester. It gets into the blood, it immediately gets saponified back to the mono-acetyl compound, which is biologically active. In turn, it gets saponified to the de-acetylated, the non-ester, which is called morphine, which is active. And so, the whole concept of giving heroin is giving a drug that is almost totally destroyed in the body at a half-life of three minutes. That means in about three or four half-lives you just don't have any more. And yet, the effect of heroin lasts for about three or four hours. So, while the half-life drops off, your biological effectiveness is a much slower slope. So don't think of the liver as being the only organ that metabolizes drugs. The two other major organs: One is blood itself, which has lots of enzymes, lots of capability; the other major organ, that often has its metabolic aspects overlooked as a tissue in the body, is the lung.

One of the earliest studies I did where I got a real surprise on that [metabolic action] were some radioactive studies I did with a compound called DOB. It's a fairly potent hallucinogenic, and we made it with radioactive bromine so we could trace it in the human body by means of an external camera. Put it in the body, injected it into the vein. And body scanning, I'll talk some about this when we get into radioactive scanning. You put a person on a table, move the table, and the person is underneath a battery of detectors. And as you're moving the detectors down the person you get a picture of the body. And where there's radio-activity in the body, you're not following the drug, you're following the radioactivity. With the radioactivity you can see the site of injection a little bit still, the tissue is a little glowing spot in the arm.

And [in this DOB study] it built up into the lung! This is a psycho-active drug that causes weird visual phenomena. Our visual center is not in the lung, but that's where it went. And it built up in the lung, the lung level began dropping off after about two or three hours, and at that point the brain level was building up. The drug happens to take three or four hours to have its effectiveness. And so suddenly I got this insight, "I wonder if the drug is really a prodrug." It's getting into the lung, the lung is doing something to it, not taking the bromine off but doing something to the molecule, which then goes on and accumulates in the brain and has its action. What is this intermediate conversion process? Unknown. Totally unknown. You can't delve into human brains and take out sec-tions and analyze them. [Laughter.] And this doesn't happen in animals. We tried it in a dog and a monkey. It doesn't happen. It's only in humans. So, we don't know the answer to that.

The point is, something is going on in the lung. It's not splitting off the bromine, because we did not find radioactive bromide in the urine. So some transformation is going on. The half-life in these various tissues is a mechanical absorption and wash-out. But the biological activity in the second half, between how much is there and what effect it has, I'll get into in the next hour because we're almost at the end of the hour.

[Directed to student] Yes.

STUDENT: The function of the pulmonary vein?

SASHA: Hoo! You can get at it. I don't think I could do it for a casual experiment.

I'll mention one thing where this is an actual application. There are two basic therapies for epilepsy. One is phenobarb, a material that is a very effective antiepileptic. The other is Dilantin, or diphenylhydantoin [aka phenytoin] which is used frequently as an antiepileptic. Both are effective. Sometimes one is more effective than another, sometimes the other is more effective. Sometimes they use a mixture. But phenobarb induces enzyme changes in the liver, and those enzymes that are induced to change affect Dilantin. And so when you have a person on Dilantin at a stable level and you introduce phenobarb as a possible alternative, that Dilantin level is going to change. And you don't know how it's going to change until you find how that liver is going to reallocate its capability according to the phenobarb it's fed. The other way applies also. So you have to all the time titrate from the blood levels when you're mixing medications because some medications affect the levels and the metabolism of others. So, blood level detection for active compounds is a routine thing in many kinds of therapies.

More Body Plumbing &
The Body's Wiring System

(or, How the Drug Acts Once It Gets There)

SASHA: As you probably got from the original outline, my hope is to start the course with sort of a picture of where drugs came from, where they go, drugs in a generic sense, how they get into the body, what they do in there, how they move around, without paying much attention to what the drug is. Then the middle portion [of the course] will be more or less devoted to specific drugs. And there is where the textbook will be a very useful ally, because I will mention each week what I'm going to talk about the next week. You can read that, what's in there, and then I can have the luxury of filling in the holes and answering specific questions without having to re-paraphrase the book. Then the last section will be, more or less, off into a kind of wilder territory. If you define drugs as those things that influence a living organism or behavior, such that I can embrace as drugs things like radioactivity and smog and pesticides and pollution and anything else I choose to, at the end I intend to go into specific aspects that influence behavior and body function but don't necessarily fall along the classic lines of drugs. So, the middle portion of the semester will be the main area where the book will be useful.

I want to do a little finishing up on the last lecture. I got into the first half of the dose-to-concentration curves, the kinetics aspect of the thing, but I want to elaborate a little bit more on it, again, with just sort of handwaving. Let's take an analogy of black ink and flowing water to go back to the body. Let's say you have a big washtub in the basement at home and you fill it full of water so the water is running out of the top at the same rate water is running into it, and to be sure of reasonably good mixing, you have a little stirrer in there. On one side of the tub is a small probe that can measure the blackness of the water, once we pour in the black ink. And you put in some nice clean white cotton. And you know that ink will stain cotton, and you also have a lightbulb. We use an ink that will stain cotton and that will degrade in the light. Something that will bleach, that will fade. And we're going to sit down here and measure the color of the water. So, you fill the tub full of water, throw in the nice white cotton, and turn on the light that does the bleaching. And then with the stirring going on, the cloth is bouncing around, the liquid is stirring around, you dump in a bottle full of black ink. And you try to get a feel for what is going to be the darkness of the color as a function of time.

Well, let's say you put enough ink in there to get to this level [drawing on board]. You have a tub of water stirring, no cloth, no light. You dump in the ink and as you stir it, you find your ink will do this kind of a curve [drawing on board] wherever your sensor is. You put the ink in here, the sensor is over here to the side, and as the ink stirs, it distributes through the water and becomes uniform. And then it stays there. So, this is, in essence, the concentration that is a function of the ink and the water. That would be exactly what the body is if you had the body with this circulating blood, with the continuous re-using of the blood, and the removal of wastes by passage through the kidneys. And you put in a little bolus, the term bolus is one I've given before. A bolus is a *gloop,* a *glop,* a single shot. You want to inject ten milliliters of stuff in the blood, you put the needle in the arm, and you go 'wham!' and all ten milliliters are in the blood. So it's coursing through the body as a real tight peak of stuff. And as it goes, it tends to diffuse and be dragged about. The edges don't move as fast as the center of a stream of blood and it gets broader and broader. And so, the bolus will give you this type of thing. In truth, the bolus is not going to diffuse, perhaps, as fast through the whole tub, as the stirring is moving water around the tub. So you might actually have this kind of a situation where you have a little bit of an overshoot/

undershoot, but eventually the sensor is seeing the average that is established by statistics.

Do the same thing in the tub, now we'll put the cloth in, the nice white cotton, and the black ink is in there. The stirring is going on. You'll find that the cloth will have an affinity for some of the ink. Ink will go into the cloth, ink will come out of the cloth, but there is an equilibrium. If there is no ink in the cloth more will go in, and if there is more in there, ink will come out. So you have a process of dyeing and undyeing the cloth. It takes a certain amount of time for the ink to get into the cloth, and once it's in, obviously the ink has been withdrawn from availability, out in the body of the fluid. But once it's there, it's static. Now, the ink is going into the cloth as fast as it's coming out. The level is lower because the cloth represents a depot for the ink that does not have any availability. It removes it from the water that's circulating. So in essence, this is the tissue distribution, the analogy to the body, and this is what is circulating.

Now we'll turn on that photo-destroying light, which tends to bleach the ink with time. And what the level will do now is go up, and then have a slightly lower level, and we'll find that it will disappear with time because it will be coming out of the cloth, in and out of the cloth, but that which is out of the cloth is being photo-destroyed. And maybe that which is in the cloth is being photo-destroyed. And overall, the curve will drop off.

At, or just after, the sudden rising of the level, an uncertain oscillation occurs. The oscillation is due to the uncertainties of circulation that make the initial opaqueness (or drug level) unstable due to the mechanical stirring (circulation paths of differing lengths). And the actual slope of the curve (half-life) will depend on 1) The size of the bucket. The larger the bucket, the longer it will take, at a given rate of water entry/exit, to rinse out the ink. But the volume of blood is a constant, as in this case the size of the bucket, and this factor does not apply in the analogy. 2) The rate of water entry/exit. The larger the input of water the shorter time it will take, with a given volume of bucket, to wash out the ink (clear the drug). The term "renal clearance" (renal means pertaining to the kidneys) is the body factor that dictates the rate of removal of drug from the blood, and is directly analogous to the volume of water flow in this analogy. But 3) The half-life, in the case of either the ink (bucket) or drug (blood), is pretty much independent of the amount of ink (drug) introduced.

That's it for the simple, unreal example. In real life there are many factors that make the picture more complex. Two of the most complicating factors involve changes in blood level of a drug due to factors other than elimination. The drug being studied will surely metabolize. These new chemicals (metabolites) can themselves have an influence on drug redistribution and drug clearance.

Yet another important variable in drug clearance, and thus the shape of our plasma half-life curve, is the acidity of the urine. In the discussion of the reabsorption of water by the kidney, mention will be made of the possible reabsorption of nicotine. This does indeed occur. Nicotine is quite basic, so if the urine is acidic, the nicotine will be in the ionized (the water-soluble) form. There will be little if any reabsorption through the tissue structure of the kidney. But if in the middle of the experiment the subject eats some food with a lot of sodium ions in it (taking a tablet of sodium ascorbate, vitamin C, for example), the urine will become relatively un-ionized, and its rate of excretion can slow down rather dramatically. This is yet another factor that makes the half-life plot something other than a nice theoretical straight line in the fifth plot of the figure.

When you go into the body with an intravenous injection, and put a bolus of a chemical into the blood, it circulates out of the blood. The blood is in continuous contact with tissue, and some of it is being sucked off into the tissue; in essence, it's being removed from the blood in a form that is no longer circulating. This is called tissue depot, or fat depot. And, of course, the liver is doing its thing, grinding away, chewing on everything that comes around it. So it's like the light that decomposes. And as it decomposes, the drug level will drop off.

Now, you have one thing in the human that may not be clear in this little model of the cloth in the tub with the light on and the stirrer going. We could have had a drain rather than an overflowing tub. But we always have the same amount of water in the tub, it's being continually replenished with new material. This is equivalent to excretion, and you will excrete as you clear the blood. The term clearance is a little technical. I don't want to get too far into it, but the concept is that you will remove material from the blood, usually at a rate that is a function of the amount of blood. Not the amount of material, but the amount of blood that goes through the liver, and will clear that amount of blood of all drugs. Or of a certain proportion, but a consistent proportion. So you define the term clearance not by the weight of the drug or the amount of the drug, but by the amount of fluid that can be cleared in a given time. The efficiency of the liver is a measure of clearance. And that is an actual removal of the drug. And so, in that case, you will have a system that will go at yet a steeper slope [referring to the previous graph]. The slope of the half-life is a function of the decomposition and the metabolism of the material. It's a function of the tissue availability and it's a function of the distribution within the blood. The metabolites can themselves have an influence on both drug redistribution and drug clearance. That is more of a moving picture of the actual metabolic system.

[Directed to student] Yes.

STUDENT: You're saying that excretion is done by the liver and not the kidneys?

SASHA: Kidneys for excretion, liver for metabolism.

STUDENT: Okay, so both are involved in it?

SASHA: Both are involved. The kidney is the water flowing out of the tub. In essence, it clears and gets rid of. The liver is the light overhead. It tends to fully decompose and convert to other things.

Let's take a specific example, let's take a drug. Let's make up numbers. Let's say that caffeine, which is one of our major drugs, sometimes thought of as a food but it's a stimulant, is half and half in the blood and in the tissue. Let's say that half of it is metabolized in a certain period of time; actually, the rate of caffeine metabolism is probably half of it is metabolized in a couple of hours. So its normal half-life would be a couple of hours. And let's say we take a bolus of caffeine, shove it into a vein in the arm and follow what's going to happen to that caffeine. It's equivalent to pouring the ink into the tub. You inject that caffeine into this arm and put a probe into the same arm. So, let's follow what happens to the 100 milligrams of caffeine going into this arm. It will go in, it will drain into the heart, it will be pumped to the lungs, it'll come back from the lungs to the left-hand side of the heart, it will then be pumped from the lower left-hand side of the heart into the artery system. Arteries go all through the body: main ones to the head, a lot of them to the gut, a lot to the kidney, a lot to tissue, a measure to the heart, and amongst the arteries are the arteries that feed this arm. The caffeine, by the way, as it starts down this arm is no longer that sharp bolus spike, it's diffused out. There has been turbulence. It goes through capillaries in the lungs, and some move more quickly than others. Some arteries are long dis-tance trips, some are short or medium trips. The artery to the heart is right there. And you'll find that a little bit of the caffeine that has gone to the heart in that portion of the artery is back in the venal circulation very quickly. Whereas the artery that goes and feeds the bottom of the foot will be a little bit later coming back in. So the caffeine level is being diffused, is being averaged, is being generalized.

Now, as that caffeine goes into this tissue [the blood] and you see it coming back in this arm, you'll suddenly pick it up. But you won't pick it up for the better part of a minute because the blood circulation takes a minute to do the whole tour, on the average, through the entire body. And, of course, when you see it here, you remember when you injected that bolus, the bolus went as a pulse of caffeine, but behind it was clean blood. And so, when you pick it up in the other arm, you'll find the surge of caffeine in it, which will drop off again because it's been washed out by clean blood. And then it gets more turbulent and you get that slight oscillation that eventually satisfies itself as a static level.

So, if the blood system were a closed system and had no access to any-thing, it was just a bunch of plumbing, and you injected 100 milligrams of caffeine (a typical dose is 100 milligrams, coffee runs seventy-five, 100, 125 milligrams per cup; tea is two-thirds of that, maybe fifty to seventy-five milligrams. Look at a bottle of Coke, sometimes it says what is in there, I think it's 150 milligrams. And they're now selling something called Jolt in which there's twice the amount of caffeine, it gives you twice the load. But you don't inject that, boom!, in the vein, unless you have really strange tastes), generally it goes in over time so you have one more diffusing agent. So if you have 100 milligrams into five liters of blood, let's make nice round numbers, this means you're going to have twenty milligrams per liter. The usual dimensions for blood in the body are milliliters, thousands of milliliters. And so you're going to have twenty micrograms per milliliter. As an ideal concentration in the blood, if the blood were merely in a closed system and didn't have access to anything outside, and there was no metabolism, no tissue distribution, no degradation whatsoever, that would be the level you would get.

Well, drugs are very rarely put into the body in a vein—*gloop!*—bolus. Because this little thing is roaring in and suddenly the concentration in the blood here is five micrograms per milliliter, and over here it's 100 microgams per milliliter. And when that hits the brain, all of a sudden, you'll have this "whack!" of a very, very potent stimulant hitting the brain and then draining away, and you're going to activate things more severely than might be wished. If it's a clinical application, you rarely want to have the impact of the injection. You would rather have the effect of the drug. And so in drug abuse, very often the drugs are injected all at once. Same idea with smoking a cigarette. You take that first inhalation, you don't seep it into the lungs, you drag it into the lungs all at once to get as big a shot as you can into the blood so it hits the brain in a shot and gives you that high, that jolt, the impact of a sudden but passing overdose.

[Directed to student] Yes.

STUDENT: So which arm would take faster to get the jolt?

SASHA: The right.

STUDENT: The right. Because they usually give medications on the left.

SASHA: The difference is small. Often the right is a stronger arm physically.

STUDENT: So for an IV drug user it would be more advantageous to go on the right, you should get more of a jolt?

SASHA: It'll be slightly faster because you have to cross to get to the superior vena cava. And usually, most of us are right-handed, the right is the stronger, the right is the more available, the more easily handled. Very often in clinical use you use both because you sometimes want to assay what you're doing without getting near the point of injection. Strangely, it's often the choice of the nurse or the doctor on the basis of what they feel is most easy. It's like having the emergency brake on the right or the left. Somehow, you're used to it there, but someone who's always driven a funny car with the brake over here will tend to reach for the brake first. You tend to know the geometry of handling what's going on in an asymmetric system by experience. There's really very little to choose between them.

[Directed to student] Yes.

STUDENT: In literature, I wonder if you can define, a lot of it gives drug percentages in kilograms, like effective dose—

SASHA: Oh, the amount administered per kilo of body weight.

STUDENT: How do you deduce that?

SASHA: That's a good aside because I want to talk a little bit about dimensions. There is some fuzziness always with these. Very often drugs are administered to a person on the basis of their body weight. So the dosage of the drug employed is stated as being a certain weight of drug (usually in micrograms or milligrams) administered per unit weight of the subject or test animal (usually in kilograms). There is a natural instinct that tells you that a heavy person should get a big dose, and a light person a small dose. Using this relationship, one can select a quantity of drug to administer which will take into account the weight of the subject. This allows an easy translation from one experimental animal to another, or from an animal to a human, for either comparative toxicities or estimates of effective levels. The function is, the reasoning is, that the more the body weight, the more the drug is going to be diluted. You have just so many receptor sites, you want to keep a given concentration to that receptor site. So you'll give more drug to a heavy person to get a comparable concentration. There's nothing really rigid and hewn in stone in this philosophy. Because after all, if you're dealing with receptor sites, most people have the same number of receptor sites regardless of their body weight. You have some people who administer a drug based on body surface. This is something that's always left me a little fuzzy because you have to calculate the surface based on body weight anyway, height and body weight. So you're at another arbitrary number.

The simplest of all is the uniform dose. Sometimes there are philosophies that you give the same amount of drug to a person as long as they are adult, or at least fully developed (regardless of height or weight), which assumes all persons have a similar number of neurons and neural receptors from infancy to old age. Or if the problem is an infection, you'd want to titrate not on the basis of how heavy the person is, but how many bacteria you have to get at (the severity of the infection) with only the toxic side effects requiring that attention be paid to body size. So all these little things will taper it.

Very often you'll hear of administrations of a drug being expressed as something like two milligrams per kilogram. A good swinging average for a typical adult, for whom you do not know either weight, size or sex, is eighty kilograms. You'll swing from sixty to 100, but about eighty kilograms. And so, if you have a two milligrams per kilogram dosage, that means you would add 160 milligrams. But again, it's capricious and often what's very strongly dictated for one drug will be totally different for another. This measure of dose, milligrams per kilogram, is a concentration. This means that for every kilogram of body weight (about two pounds) there will be administered one milligram of drug. The two pounds image is easily visualized: A couple of bags of unground coffee beans is a kilo of coffee beans. But how can one find an analogy for a milligram? A coffee bean itself may be 150 milligrams, and it is difficult to try to visualize a 150th of a coffee bean. Clip your fingernails. A clipping weighs about five milligrams. Clip the clipping into fifths, and take one of these. A milligram of fingernail. Many drugs are potent and even lethal at a thousandth of this concentration (about the weight of the small fragment of a fingernail clipping in not two pounds of coffee beans, but two tons of coffee beans).

Generally, most animal studies are made in milligrams per kilo of body weight. But here, I'm assaying the amount of drug. This 100 milligrams going into an eighty kilograms person would be just over one milligram per kilogram. All of the blood levels and urine levels are also measures of concentration, and are expressed in similar terms. With these, as with most fluids, the denominator unit is volume rather than weight. In the five liter blood pool it would be 100 milligrams in that five liters, or twenty micrograms per milliliter. Thus, if one milligram of drug is dissolved in one kilogram of urine, one correctly assumes a urinary specific gravity (density) of unity (meaning one) and speaks of the concentration of one milligram per liter. Since most assays use much smaller volumes of urine, this specific concentration may be written by dividing both dimensions by 1,000, i.e., one microgram per milliliter. A valuable family of prefixes to have reference to is the group that signifies relative multiples of one thousand. Each successive line below is 1,000 times less than the previous one, except for those that straddle unity. There, each multiple is a factor of ten. The upper and lower extremes are almost never seen, but are given here for fun!

One additional, rather old-fashioned measure of concentration is occasionally encountered. This is the measurement of concentrations in weight percent. A number of years ago it was the standard format used in clinical chemistry, and the weight per milliliter term was largely

exa-	a quintillion	1,000,000,000,000,000,000.
peta-	a quadrillion	1,000,000,000,000,000.
tera-	a trillion	1,000,000,000,000.
giga-	a billion	1,000,000,000.
mega-	a million	1,000,000.
kilo-	a thousand	1,000.
hecto-	a hundred	100.
deca-	ten	10.
unity	one	1.
deci-	a tenth	.1
centi-	a hundredth	.01
milli-	a thousandth	.001
micro-	a millionth	.000,001
nano-	a billionth	.000,000,001
pico-	a trillionth	.000,000,000,001
femto-	a quadrillionth	.000,000,000,000,001
atto-	a quintillionth	.000,000,000,000,000,001

restricted to toxicology. Today this weight percent term is found only in the measurement of blood alcohol. A blood value of 0.1 g% means a tenth of a gram contained in 100 grams (or 100 milliliters) of blood, and is synonymous with one milligram per milliliter (0.10 g% = 1 milligram per milliliter). More of this will be discussed during the lecture on alcohol.

Let's go into the second thing where it goes off into the tissue, back to the analogy of having the white cloth in the washtub with the black ink going in. Let's say that the caffeine is distributed half into tissue and half in blood. Let's say its partition is one. Partition coefficient is a measure of the ratio of concentrations between two separate systems, very often water and oil. In this case it would be blood and tissue, or blood and fat. This is a partition coefficient where you have equal concentrations, you have a partition coefficient of one. So you would have not twenty micrograms per milliliter, with 50 percent tissue you're going to have ten micrograms per milliliter, because ten micrograms is in the blood and ten micrograms is in the tissue. Actually, there's more tissue and fat than blood, but let's keep this thing in a sort of a round number basis. Therefore, if only ten micrograms is in a milliliter of blood, you speak of the volume of distribution as being ten liters because you have five liters of blood, but the concentration, you see, is as if you had had ten liters of blood. So nothing can have a greater concentration than all of it being in blood; but if there's any tissue extraction, any extraction out of blood, binding in a way that it's not available to blood, like protein binding, then the level in the blood will be lower because of that. And after all, it's the level in the blood that makes it effective, usually at the receptor site that the blood perfuses.

Let me give another illustration of this concept of dose, volume of distribution, kinetics, and excretion with an actual drug, and use it as an introduction to pharmacodynamics. I used caffeine as an example, but now let me use the real values associated with the drug nicotine. The actual details that concern nicotine and tobacco will not be discussed until the tenth lecture.

Nicotine, the stimulant and pleasurable component of tobacco, is about 1 or 2 percent of the weight of tobacco, so with a cigarette that weighs a gram there is perhaps ten or twenty milligrams. During the volatilization that accompanies combustion during smoking, perhaps nine-tenths of it is burnt, and only one-tenth is delivered to the smoker. Assuming all of that load is absorbed and retained, a dose of nicotine will be about two milligrams. So, after a morning of five cigarettes, the subject has taken some ten milligrams on board.

The volume of distribution of nicotine is about 200 liters, and the blood half-life is perhaps two hours. What can one expect in the way of body fluid levels?

Since there are, say, five liters of blood, and some 200 liters of apparent distribution, thirty-nine-fortieths of the nicotine is *not* in the blood, but in tissue. The amount actually in blood then is one-fortieth of the ten milligram dose (250 micrograms) and it will actually be smaller, since by the time of the fifth cigarette some of the nicotine of the first has already been disposed of (metabolism and excretion). So maybe 200 micrograms will be dissolved in five liters, which is a concentration of forty nanograms per milliliter. The concentration of the major metabolite, cotinine, is about ten times this level.

Nicotine and cotinine are both excreted, each accounting for about 15 percent of the total initial nicotine. Over the course of a day a person will be putting out maybe a couple of liters of urine, and with 15 percent of the original ten micrograms of nicotine that had been absorbed being excreted unchanged, one would expect urinary levels of maybe one-point-five milligrams per one-point-five liters, or 1,000 nanograms per liter. Cotinine is seen at about the same level.

PROPOSED ROUTES FOR THE METABOLISM OF NICOTINE

Why isn't the inky water overflowing our tub of exactly the same darkness as the inky water inside of it? Or why isn't the urine level of nicotine (1,000 nanograms per milliliter) exactly the same as the blood level (forty nanograms per milliliter)? It is not the same, simply because the kidneys are not a valve for the release of blood, but rather are a get-rid-der of things dissolved, with a different mechanism for the reclaiming of water. The water (free of nicotine) is reabsorbed and reused. In the tub-overflow example, fresh water was used continuously, and old water discarded. In real life, there is no "blood faucet" and everything is recycled. Even a little of the nicotine!

And the heart of all this body-level measurement is simply to evaluate, from those things that can be measured, whether a drug does or does not have an effect. This latter property is a much more difficult thing to measure. This is the meaning of the pharmacodynamics term of this lecture: the relationship between drug concentrations and drug effects.

Let's say a drug is being metabolized. The metabolism occurs over a period of time. Let's say each pass in the liver constitutes a removal of 50 percent of the drug. That means 50 percent will be removed per minute because the blood will make it through the liver about once a minute. Almost a great deal, about a third of the blood circulating in the body, makes it through the liver. Some of it goes to tissue and goes directly back into veins. Some of it goes to the brain and goes directly back into veins. But about a third of the blood goes to the gut, or to the liver by means of the gut, or by means of aortic arteries. And every minute that amount will be taken off. So this number will drop by about, let's say, 50 percent per minute of that blood which is going through the liver, which is about a third of the blood, so about 20 percent of this will drop per minute, as the half-life, due to metabolism.

Now, if you're looking at the drug itself, once it metabolizes it's a different material. You will have a certain amount that will also go to the kidneys and be extracted by the kidneys in the renal filtration process and be put into the bladder. Of course, the metabolites themselves would not count in the concentration of the drug. But occasionally you find such things as metabolites that influence the function of the kidney, or influence the function of the liver, or influence the function of action of a drug. Not to say that the metabolite is the active species, but the metabolite can, in turn, induce other changes that will affect the concentration of the drug. That's why when you plot a half-life curve of drug from blood, you do not get a nice beautiful mathematical straight line. You get a weird overshoot and then a strange slope-y thing dealing with many, many compartments and many, many fates of a drug. Let's take that caffeine injection example and do one more scenic tour. Let's do it by taking it orally. Oh, let me go back. One thing about the bolus. The way you bypass the bolus effect, quickly getting the intense over-concentration, is to inject using what's called an infusion. This is where you will put the drug slowly into the arm or have a drip in the arm and add the drug slowly to the drip where fluid is continually going in. With 100 milliliters over the course of half an hour you could add to the blood at a rate that the overall concentration will build up very, very slowly. And as it builds up, of course, the metabolism tends to drag it down, but you do not have this overshoot phenomena that you wish to avoid.

In the recreational use of drugs, that overshoot is exactly one of the incentives to use the drug. When you watch someone or read about the injection of amphetamine or methamphetamine, you will have a person who has the chemical or the drug in a syringe and the syringe is in the arm and the person has pulled blood in the syringe to mix the drug into the blood. Or this is done with heroin, in which it's mixed and mixed and played with as part of the ritual. Needle orientation is unbelievable. The direction is not to the drug so much as to the needle itself and its use. It becomes part of the mystique of drug use. And then, they whack it in, pull off the tourniquet, and this big charge of the drug goes and hits the brain, oh wow. And then you'll have a person who will fall back on their back.

You have much of this in the use of snuffs. I've seen movies showing the use of the *paricá, vilca,* and the various *yage* snuffs in South America where they're taken up the nose. This is where you get it right into the tissue and it's absorbed right away. You'll have a couple of people facing one another in a cooperative way, and they'll get a couple of hollow bird bones, one end in the mouth of one, the other up the nostril of the other, and in a given moment they'll both blow as hard as they can go, and the drug goes up the appropriate nostril. Watching this in a film is amazing, you sort of stand back and take the picture. You will find that after the blowing of the bird bone they usually end up on the ground; they both go over on their backs—right away. Whereas if you diffuse it slowly, you'll have it creep up on that impact, but that is often not what is wanted. What is wanted is the suddenness, the impact of it. You want that bolus, you want it to get there as fast and as hard as it can. I mean, why is the use of nicotine gum not as satisfactory as smoking? It doesn't build up to quite as high of a level. It's metabolized more directly because a lot of it goes down into the stomach and goes to the liver. But with smoking you get an impact of a sudden burst into the lung which is carried in a *wham!*—right into the brain. And you have that very sharp, "Oh wow, I'm awake" kind of experience.

I'll get into this more in the discussion on tobacco, but I was in a discussion in a seminar on Monday in which a person was making a study to find out if there is an effect on the metabolism of nicotine that comes from blood sugar or from eating. They had run a survey in a small group asking if people had to give up all but one of their cigarettes, which one would they maintain? Which was the last that they would give up from smoking. And most people said the one after dinner. And so why would a person choose that? Was there something about food going in that decreased the nicotine level and made a special call on it? Not necessarily, but the question can be addressed by giving nicotine infusions to build up a certain level and then supplying the nicotine continuously to maintain that level, and then give 800 calories to eat in ten minutes and see if that level drops. If it does, then there's an argument that there is an influence of food, blood sugar, portal circulation, whatever, that would cause the nicotine to drop and would send the signal to the body: "A smoke would be very nice."

Well, we had a rolling discussion after that. I'm a person who had smoked for a number of decades. I contributed my two cents worth that I think the nicest is the first one in the morning. The last one I would want to give up would be the first of the day, where you are really down on nicotine and it has the greatest impact. And another ex-smoker climbed right onto that and agreed with me. Well, it turned out in the experiment there was not an appreciable drop in nicotine in the function of food eating. Interestingly, the irritation of not being able to smoke was not nearly as great as the irritation of having to eat the food in ten minutes, apparently 800 calories. When you're locked into a ten-minute thing, of course coupled with the fact that you've got lines up both arms and you're strapped into an experiment (they get free meals and free lodging for three days and some extra pin money, but still), it is not a pleasant thing to have blood drawn every hour or every half hour, especially when you don't get a positive result out of the experiment. No, I think that rush, that impact, is one component that is a very favorable one that encourages continued drug use.

So if you were to take that same 100 milligrams of caffeine orally, it's much slower to come on because for one, it lands in the stomach and caffeine is not absorbed from the stomach. It has to make its way, take its place in line, with whatever other food is in the stomach. Often coffee is taken after a meal, and after a meal the stomach is especially enriched with food. It makes its way into the small bowel from where it is absorbed and transported by this very thoroughly efficient system into the liver. Whereas intravenously, it will reach the brain before touching the liver. When you take it orally, it makes the liver before it touches the brain. And the liver, like a toll taker, takes its 30 percent, or whatever percentage, of that caffeine before it releases it into the circulation. Then it goes into circulation, and every time it comes around, the liver takes out its percentage again. It is a continual metabolizer. But the "first pass," the term used in pharmacology and pharmacokinetics, is to the liver and the liver therefore has a chance to get at, and will take and metabolize, a certain quantity of the material. Intravenous will avoid the first pass of the metabolism, at least in the liver, and oral will not. You find your levels will never get as high, in the case of a drug such as caffeine, if it goes in by mouth first.

Okay, that's kind of an illustration of kinetics. It's easy to say that you can tell how much influence a drug has by how much the blood level is. This is not a very sound generality. Looking again at the last of the graphs, there is a relationship portrayed between blood level and time. This clearly shows how body levels will vary, and inevitably decrease, as a function of the time that has elapsed since the drug had been taken. But how can you tell from this graph when the person will be feeling the effects of the drug?

Questions related to this are commonly asked in both medical and legal situations. "What is the level of Dilantin that is needed to guard against convulsion?" "Was so-and-so under the influence of phencyclidine when he shot his wife?" You are handed a blood sample and asked to deduce behavior from some microgram per milliliter number.

The only drug I am aware of that has a rigid assignment of behavior to blood level is alcohol. And that has been dictated by law rather than clinical observation. If you have ethanol in your blood at a concentration of 0.10 g%, you are under the influence of alcohol, in California, when it comes to the matter of driving a car. The number is legally hewn in granite, and will be evaluated in court, if need be, later.

But from the clinical point of view, there is no fixed and dependable relationship between blood level and drug effectiveness for any drug, not even alcohol. This person is different from that person. Today is different than yesterday. The second time is different than the first time. Such-and-such a concentration is different going up than it is going down. After a while at this level, the effects began to change. These are just a few of the variables that make this a difficult question.

The question comes up quite often in legal areas: was the person under the influence at the time they were driving the car into the tree and were killed? You'll go into a dead person, you take out a bunch of blood, a heart puncture is the easiest way of getting a quantity, and you measure the amount of phenobarb in it, or the amount of alcohol, or the amount of caffeine, or whatever drugs you see, and you say, "The level was such and such. Therefore, since the active level is such and such, it's above the active level, so they were under the influence." Not so. It is easy to simply draw a line across the graph at some blood level, as indicated by the horizontal line A- - - -A. Then one could say that everything that lies above it, like the visible portion of the iceberg, represents drug action. And everything that lies below it represents an inactive level of the drug. Then, the duration of action would be the time that the iceberg tip emerged to the time it disappears again. Easy, and totally unreal. Because you do not know if the thing was on the way up or on the way down. If, indeed, the onset of action of a drug were to be pinned to a given blood concentration on the way up the curve, the action has usually ceased before that same concentration has been reached again on the way down the curve. Very often you'll find, in a very global, hand-waving sense, if you have a course of a drug (let's take the more characteristic curve) like this, you will find that the active level will be from here to there [demonstrating with graph]. Namely, its first activity is shown at such and such a level on the way up. That's when it gets to the brain, if it's a centrally active thing, when it gets to whatever it's being active on.

But it's not the same place going down that the activity ceases. The activity usually ceases quite a bit higher. Partly because the receptor site has been perfused and is tired. Tired is an unscientific word but that is the idea. It's been used for a while and it is not as responsive. You have what's called short term tolerance where the receptor is being continually provided an impulse. Something is depleted or something becomes less responsive. You take a person and give them a sedative hypnotic dose of something like secobarbital. Here's where they drop to sleep, here's where they wake up, and here you have the blood level where they're sound asleep, but they're quite awake. So if they went into the tree at this point, you'll say, "Clearly, they were under the influence of the drug. They were a bad driver. They had secobarb in them and it caused a certain amount of things." You take the blood level at this point [indicating on the graph] and there's no appreciable inhibition of driving skill and yet the levels are the same. And yet there is the "hangover" aspect that reflects just this depletion, possibly nerve irritation, possibly transmitter exhaustion, that might not be specifically a "drug action" but which honestly reflects some impairment that is a consequence of drug use. We will talk quite a bit more about this problem with the alcohol situation later in the course.

[Directed to student] Yes.

STUDENT: How does hangover fit into this?

SASHA: Hangover from alcohol, from the barbiturates, probably as much as anything is disruption of the nerve integrity. Alcohol has a property of going into the body and being metabolized at a slow but constant rate. But it is stored in the nerve sheath, in the linings of the nerves, and you have a continual irritation even though there is not alcohol there, the nerve has been irritated and must repair. So, a lot of that is not due to the presence of the drug, but is due to presence of transient damage or insult to the nervous system.

[Directed to student] Yes.

STUDENT: Is there any way to flush the alcohol out of the neural sheath?

SASHA: Not that I know of.

STUDENT: They say, you know, drink a lot of water and stuff like that.

SASHA: Oh, that's good for getting alcohol out. No, it's not even good for getting alcohol out. There are as many recipes for getting rid of a hangover as there are for starting a Grignard reaction, so to speak, [laughter from class] from the chemical point of view. Everyone has a way of doing it. I'm going to spend a whole hour on alcohol, I'll get into much of it. The alcohol is debilitating in a continuous way. Alcohol does not appear to follow this curve, but it does. You can overshoot the amount initially, and get drunk more readily. Now, that's not totally fair because alcohol is absorbed from the stomach to some extent, it's one of the few drugs that actually being neutral can be absorbed from the acidic medium.

For example, you take absorption from the lung. One trick we used to do in college when I was young and in wilder days, and you're working at about two o'clock in the morning and you're tired of waiting for some chemistry to be completed or for something to happen, and you're bored, what we'd do is run hot water into the sink and really get rolling hot water with steam. As much hot water as you could. Then we'd take a liter of ethanol and throw it in the sink and stick our heads down in there and inhale through our mouths a couple of times. Oh! You are just right there. Into the lungs, into the brain! None of this first-pass metabolism. No first-pass that hits the liver. You're very quickly roaring drunk, and then it diffuses on through the blood system and gets more or less averaged because you're getting a bolus to the brain. Once it gets averaged, the average level is too low to be intoxicating and you're sober about ten, twenty, thirty seconds later, but a little less bored. It's one of these real intense intoxications that, once it's averaged out, is not there.

So, the blood level has not yet been averaged out and you're looking at a blood-level concentration that has gone to the brain because it went directly into the pulmonary system. Remember, the blood that goes into the lung and goes back to the heart is pumped arterially out right there. From the lung to the brain, a few seconds. From the vein to the brain, quite a few seconds. From the muscle to the brain, quite a few more seconds for yet other reasons. From the stomach to the brain or muscle to the brain may actually take minutes; from the stomach to the brain, many minutes, depending on other things.

The opposite can be seen where there is an intention to slow down the process as much as possible. For example, one of the tricks that's used in modifying tissue absorption in injecting tissues, you'll find that you'll do this when injecting an anesthetic like Xylocaine into the gums in dental work, is that you'll put the drug in there, but you'll also put in a little norepinephrine (which we're going to get to if we have any time left at the end of the hour for neurotransmitters), which tends to constrict capillaries. It tends to make capillaries smaller and harder to get blood into and out of. Closing down capillary circulation impedes the process of sweeping the local anesthetic out, to its eventual inactivation in the liver. What you always want in tentative medicine is to use as small an amount of drug as is effective. In running an anesthetic into the tissue, if you're going to do some skin surgery or something topical, you can use a smaller, and hence safer, amount of anesthetic, less than you would think you would do, by mixing it with a drug that will close down the capillary bed so it's not carried away as rapidly.

So you add norepinephrine to restrict circulation, then add a smaller amount of drug which stays there longer because it can't get away. So, by using mixed drugs you sometimes can minimize the exposure of the person to the drug. A general premise in medicine that should always be observed and often is, use as small an amount of drug as you can get away with because you are always having a competing reaction of toxicity, a threat to the body. But a local anesthetic like Xylocaine or Benzocaine; these are local anesthetics but they are also central anesthetics. Think for a moment if you were to inject Xylocaine, which is used in tissue to deaden the tissue, what would happen if you injected it intravenously? Well, you don't have pain receptors down the vein, but you do have a general anesthetic response from anesthetics getting to the brain; it will tend to cause a person to go into a coma-like sleep.

But the toxic threshold, that difference that I mentioned between the therapeutic index of intravenous procaine or Xylocaine and a lethal or a damaging level, is very, very narrow. There has been intravenous use of procaine or Xylocaine. Lidocaine is another matter. It's used in certain heart insufficiencies in emergency surgery and in emergency care. But the usual anesthetics usually do not wish to be injected into the vein, they are used to surround the tissue around the nerve center. In the use of a spinal block, you're not injecting the anesthetic into the spinal fluid, you're bathing the synapses around that part of the spine to deaden the transmission of impulses that are coming out at that point. You do not go in. It's an error, and tends to be a very serious error to extend the needle a little bit too far in a spinal block, and actually end up in the spinal column in a spinal block.

Okay. That is an intentionally generalized picture of the drugs getting into the body, being distributed about the body, getting to sites for their action, and eventually being removed from the body, partly by excretion, partially by temporary storage, and partly by biotransformation. What I want to do is let that go for a while. In specific drugs, we'll get back to this image again. We've talked a bit about the plumbing of the body. Next, I want to talk about the wiring of the body, to draw a picture of what drugs do when they get to their active sites in the body. By the way, the handouts are strictly for the music. If you are a chemistry student you will make sense out of them but I just wanted to have them for the music of the neurotransmitters and the nerves themselves.

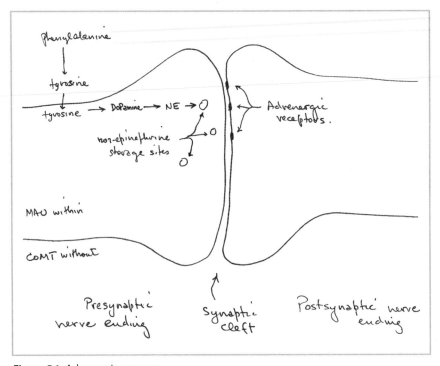

Figure 5.1: Adrenergic synapse

---

1 Handout chem structures, 4 pages starting on p.191

Two very useful words in discussing the nervous system are the words afferent and efferent. And, as with so many things, they're in alphabetical order. "A" before "E," afferent before efferent. Afferent is the direction toward the central nervous system. What is the central nervous system? The brain, the inside of the spinal column, inside of the skull, those things that are protected from the outside by two pretty good defenses. Physically, there is the cranium, the skull, which can ward off some pretty hard mechanical knocks. And biochemically there is another defense known as the blood-brain barrier which is equally effective in warding off most drug knocks. It's like an invisible Berlin Wall, so to speak, except it's not a pile of bone and impermeable tissue, it's an effective way of keeping charged things, and most things, out of the brain. The body has a barrier to materials being in the body and making it into the brain. There are tricks to bypass the barrier. Some of the nutrients that are mandatory for normal brain function (such as glucose and certain amino acids) have active transport mechanisms. It is as if there were a hand that could reach through the blood-brain barrier and literally drag a hydrated glucose molecule out of blood and into the brain tissue where it is needed. And there are pharmacological tricks taking advantage of the fact that the brain is itself a metabolic organ, and that the blood-brain barrier works both ways. A neutral drug may be administered to a patient, and after passively diffusing into the brain, it is metabolized into a highly charged and pharmacologically active thing which cannot get out again. This barrier is not perfect. There are small holes in it here and there. In the presence of a strong magnetic field, it loses some of its integrity. There are also things the barrier is not efficient at in certain situations and in certain pathologies, but in general, the barrier keeps the outside world outside, and the inside world inside. That outside world is the periphery, or the peripheral nervous system (PNS). The inside world is the central nervous system (CNS).

Anything that comes from outside toward the brain, neurons that feed signals, inputs, what have you, are called afferent signals. This provides the only awareness that the brain, the person, has of the outside world.

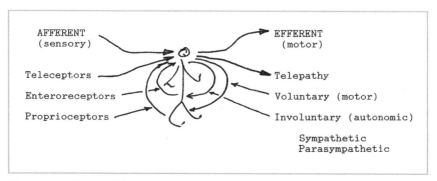

Figure 5.2: Afferent / Efferent

Nerves and nerve systems going out of the brain are called efferent signals. And we'll get into talking about such things as afferent anesthetics, drugs that do not make you numb, do not make you unconscious, do not make you free of consciousness, but block signals from getting to the brain. We'll talk about things like PCP and ketamine, and such drugs that are primarily effective by getting in the way of the input signals. Not how they're registered and not what they do, or what they promote, but keeping them from getting there in the first place. So in the afferent world, let's take the average person, afferents are things that come in from out there.

"Tele" is a very common prefix meaning "at a distance." You have the counter, "proximo," meaning nearby or adjacent. You'll find in chemistry the term "T" or "P" depending on whether you're dealing with either of two atoms, one being further away and one being closer. There are some systems in which you cannot give numberings to atoms easily, because you substitute one of them and in the other the numbering changes around. Look at the weird numbering of something like histamine sometime. The nitrogens in that ring are completely bizarre in number because they don't stay still. Once you substitute one, then you have to count a different way and the number changes. So you use the term "tele" or "proximo" and you find a little Greek *tau* (τ) or *pi* (π) to indicate which nitrogen you're talking about. "Telephone," to talk at a distance. "Telescope," to see at a distance. "Telemetry," to measure at a distance. "Telepathy," to feel or suffer at a distance. And you have the general term of "teleceptors," receptors or nerves that receive signals from a distance, the eye, the ear, smell. Things that come from outside the body directly into the body

You have a whole family of exteroceptors which will touch the body and be recorded into the head. Touching, pain, heat, cold, all require some intimate interaction between the offending agent and the offending party (in the case of irritation or pain). All these inputs are outside the body, but they require actual contact to be made with the body. In a sense, you can be picky and say, "Well, the photons from yonder tree which impacted upon the rods and cones of my retina and went to my occiput really touched my body." Or that the molecules of the smell factor of a dead skunk outdoors had to make it into the nose and physically make contact with some receptor there. Okay, picky. But in general, you're dealing as if you're not really attached to it and it is something that came to you. The smell could well have come from a distance; but, you weren't aware of it until it got into certain receptor sites in the nose. So you have the things that come from a distance, but actually require contact with the body, exteroceptors, or the cutaneous sense, the sense of the skin.

You have visceral senses, things that come from within the body. The proprioceptors are from the muscle itself. A good example of proprioception is the knowledge of where your body is, when you're not paying attention to where your body is. Without looking, keep your eyes up toward me, and each ask yourself which way is your right toe pointing. You know somehow. Without moving, without looking, you know your right toe is pointing that-a-way. You can almost point, my toe is pointing that way, yours may be pointing that way or down or up. Without moving it you know where your muscle is. That is proprioception. That's where you kind of know, in the dark, you reach over and you go right to the light switch. You have a continuous muscle knowledge of where you are. You know the light switch is over there when you come into a dark room and you feel around. Often you have a good knowledge of where your hand is without looking at your hand. Afferent still, it's coming into the brain from the body. This is proprioception.

Interoceptives are signals that go to the body, a lot of them, from the gut or from the heart, and go into the brain. All these are afferent proprioceptive inputs. You get a little bit too much oxygen. The pH in your blood goes a little bit off center for some reason. The glucose levels drop. Your whole body is aware of the fact that your glucose level is down and you're not saying "Hey, my blood sugar is down." You may say, "I'm not feeling quite myself" or "I'm hungry" or "There is something I'm aware of that correlates with low blood sugar." But you have learned the term "low blood sugar" from that feeling, and in truth, you've gotten that feedback from anyone who measures your blood sugar and says it goes down like that, you need this or that for insulin levels or sugar levels or what have you. When the blood glucose level rises after a bit of sugar-rich food, you are not consciously aware of this, but the brain gets the signal and puts out the message to the pancreas to slip a little more insulin into the blood. All these are signals that go to the brain (afferent) and provoke compensatory responses (efferent) to correct or accommodate the unbalance. You'll also find that they correlate with blood pressure. Your blood pressure gets a little bit out of normal, the signal goes to the brain which does something to bring it back to normal. All these feedback loops going to the brain as a signal are afferent signals. Coming from the brain with a signal is an efferent signal.

And what do you have going back? You have things going back to muscle. The CNS has direct communication with the peripheral nervous system (PNS) and that really has two big branches. One branch is voluntary, that is, you're actually moving to do something. This is usually thought of as a motor system (intentional control of muscles). The other is the involuntary branch that is nominally outside the control of a conscious person, often referred to as the autonomic nervous system (ANS). It is possible there is a third efferent pathway that operates at a distance from the body, this has been called telepathy, and it is difficult to evaluate experimentally.

When you're dictating your hand to go into a fist, that whole act results from the peripheral voluntary nerve system. But not all motor responses are so patently voluntary. You're sitting there and all of a sudden you go "gluh," like that, or you're lying in bed and you're almost sound asleep and you go "twang!" and your body jolts as if you've been hit with something electrical. You may even think there was a sound that accompanied it. I think everyone's gone through that. That's involuntary. There are responses, like the kick of the foot that follows the hammer tap evoking the patella reflex, you're dealing with some loop that's outside of the voluntary system, but it's still a motor system. So not all motors are dictated totally by this voluntary system, but the term has become a lumped together term for the command from the body through efferent, going out, signals to the muscles.

The second branch is the autonomic nervous system. Drugs act on the autonomic nervous system. It is commonly called the involuntary nervous system, and that is also not a correct term. The textbooks say that it is out of your control. Nonsense. It is certainly within your control. You may not be able to just sit down and say, "I am going to will my blood pressure to go down," and yet there are people who have learned ways of, one way or another, consciously influencing their blood pressure to go down. How can you voluntarily do something that's part of an involuntary system? It can be done. I had this very, very neat process to find ways of controlling individual muscle systems. Since there is no conscious awareness of the actual blood pressure, there is the need of some measure, some feedback, that can let you know consciously if you are doing the right thing. You put an electrode in a muscle and the electrode, when the muscle contracts, will pick up an electrical signal. (We'll get into something of the nerve operations in a few minutes.) And you see the needle go across, and you say, "Practice contracting that particular muscle bundle." Individual muscles are rarely independent. They work in little clusters, and they're often activated in little clusters. But you get into one of these bundles of muscles and you say, "Well, make it go." Well, you try to isolate like that and instead you make everything go. It's hard to make it go without making your hand move. And you'll find, "Huh. There it went. I wonder how I did that?" Try this, try to focus on that, it's like guided imagery in the cases and approaches to certain types of illnesses. And then it moves again.

My first wife had a stroke, a cerebral vascular accident (CVA), that was the outcome of a surge of blood pressure. She had high blood pressure running, at the time of the stroke and shortly thereafter, a systolic of around 300 millimeters, which is more than most veins can carry, or most arteries can carry. It was a very, very high level. She had an aneurysm. It was in the center, so it affected the sides of the body in a strange way. She was on medication afterwards to lower her blood pressure and to quiet stress. They used a diuretic to help clear out fluids and they were using reserpine as a de-stressing agent in the hope her blood pressure would go down. Well, diuretics tend to do strange things in the body such as nausea and muscular cramping, and believe me, reserpine can put you in a terribly depressive mood. And so I said, "Let's try getting your blood pressure down by your doing it." "Well, I can't do it." "Nonsense! Let's try it." And so, what I did was I hooked up a cuff on her, left it there, and ran her blood pressure. And, every now and then, I ran the blood pressure on her. It was running at that point about 210, it had to go down from that point. The terms in blood pressure are systolic and diastolic. Systolic is the peak pressure going out of the heart. It is produced by the contraction of the heart, first of the atricular top half and then of the ventricular bottom half of the heart. Diastolic is the pressure at the bottom of the pulse when the pressure is at its lowest. The surge is systolic, and the relaxation of the surge is diastolic. Systolic is the breaking of pipes, but diastolic is the work that the heart has to push against. So they both have different significances.

The old rule of thumb (it's kind of a loose one) is your systolic should be something like 100 plus your age. Obviously, it's not always, but it still gives you a good feeling. At your level, a good systolic would be maybe 120, 110. Lower? No harm done unless you suddenly run out of sufficient blood to get sufficient air to the brain to keep you conscious, at which point you faint. That's okay. Unless you stay without enough blood, at which point your brain would undergo brain damage. The diastolic is the labor the heart has to work against, and it should be low and stay low for your body, just as a matter of keeping the heart without having it work too hard. One-twenty over eighty, lovely. If the diastolic gets much above ninety, it's been said anything above ninety needs medication. No, I don't think there is anything that demands medication. Anything above ninety means attend to it and bring it down.

Autonomic Nervous System	Sympathetic	Agonist (Sympathomimetic)
		Antagonist (Sympathomimetic)
	Parasympathetic	Agonist (Parasympathomimetic)
		Antagonist (Parasympathomimetic)

The autonomic nervous system is a system in the body that keeps all the internal variables, all of the things that are under neurological control, in balance. There are exceptions, but in this case assume the generality applies. These correct levels are all found and maintained by the balance of two separate and opposite neurological forces, called the sympathetic and the parasympathetic divisions of the autonomic nervous system. With the autonomic nervous system you are looking at, say, the dilation of the eye, or the retention of urine in the bladder, or the blood pressure, the baroreceptors, whatever. Everything is in balance. The sympathetic branch of the autonomic system is trying to pull it this way, the parasympathetic is trying to pull it the other way. The sympathetic makes you alert, makes you be able to move fast, fight off grizzly bears who are after your lunch, run if you see hazard on the street, defend against attack. The drive is directed towards depression of secretions, contraction of blood vessels, increased heart activity, dilation of pupils, loss of bladder control, pyloerection, anything that would promote defensive capability. The parasympathetic system is what digests food, wallows, goes to sleep, makes love, all the things that are not immediately threatening and the drive is to counteract all of the sympathetic symptoms mentioned above. And all these things are kept in balance.

Consider the eye. It's a good example. In fact, it's a superb example and can be used to portray how various drugs can interfere with the integrity of the autonomic nervous system. The eye has a pupil. Here's the eye, here's a pupil. That's a ridiculous way of drawing it. [Laughter from class.] Okay. How large is the pupil? Well, sometimes the pupil's quite large. It's called mydriasis, where the pupil gets quite large, a lot of light gets in, but your vision gets fuzzy. You go sometimes into a medical treatment where they will inject a parasympatholytic agent into the eye and your pupil goes like that [demonstrating effect] and you're blind. The term belladonna ("marvelous, beautiful woman") came out of the use of this particular vasodilator, eye dilator, in Europe because the large black eyes were considered a thing of beauty. The use of atropine or scopolamine was a standard cosmetic agent for centuries. So, you had these beautiful ladies stumbling around virtually blind, suffering dry mouth, dizziness and quite a bit of confusion, but their eyes sparkled, and for that they were willing to suffer all side-effects. The name has also been given to the plant source of those drugs (the deadly nightshade, *Atropa belladonna*).

On the other hand, there are certain chemicals that will cause the pupil to go down to a pinpoint. How do these chemicals do it? The eye is a superb example of this push and pull, give and take, of the sympathetic and parasympathetic nervous systems. The pupil, for example, is the size it is because one whole system is trying to make it larger and one whole system is trying to make it smaller. It is an opaque, flat tissue that is interlaced with two sets of muscles. You have radial muscles that come out from the pupil, oriented outwards much like the spokes of a wheel, that are trying to contract. Muscles don't uncontract. They can be flaccid, but they can't push apart. Muscles are either where they are or they're closer together. And when they're not activated, they are where they are. So you can't enervate a muscle to make it longer or more flaccid. It is either flaccid or it's contracted. So here are these radial muscles that are trying to expand that pupil, make it larger. You have another whole set of muscles that are called sphincter muscles. They go around the pupil in circles of an ever-increasing diameter; all concentric with the pupil itself. And as they are activated they want to contract and make the pupil smaller. And so your pupil is what your pupil is by a balance between the radial muscles pulling out and the sphincter muscles trying to contract in. It is not that one or the other of these systems is at work. They both work all the time, and the pupillary size is determined by which system is the more dominant at any one time.

This is an excellent place to introduce the four basic classes of drugs that interact with the autonomic nervous system. There are drugs that imitate the sympathetic nervous system (sympathetic agonists, sympathomimetic agents) and those that inhibit it (sympathetic antagonists, sympatholytic agents). Similarly, there are drugs that imitate the parasympathetic nervous system (parasympathetic agonists, parasympathomimetic agents) and those that inhibit it (parasympathetic antagonists, parasympatholytic agents).

Many of the drugs that act primarily on the central nervous system can be classified by what kind of side-effects they have in their actions peripherally on the autonomic nervous system. This structure can be outlined:

The radial muscles of the pupil are sympathetic. The stimulants are sympathetic. Let me get two more terms down in here: "mimetic" and "lytic." Mimetic means something from mimesis in Greek, something that imitates. "Lys" means to cut, disrupt, decompose, disintegrate. You can have drugs that are sympathomimetic. They will activate the sympathetic nervous system. Amphetamine, for example, activates so that you're wired up, you're all wound up there, your hair is standing on end, you get that little funny feeling of paraesthesia, your eyes are dilated, you can't go to sleep, you don't particularly want to eat.

For specific examples of these agents, vis-a-vis eye responses, there are:

Class	Drug	Pupillary Response (CNS action)
Sympathomimetic (imitate or stimulate)	Amphetamine	Dilation (direct action) [stimulation, loss of appetite, sleeplessness]
Sympatholytic (disrupt or interfere with)	Clonidine	Minor constriction [nasal stuffiness, heart slowing, hypotension]
Parasympathomimetic (imitate or stimulate)	Physostigmine	Pin-point pupils (direct action) [muscular weakness, nausea, vomiting, spasms]
Parasympatholytic (disrupt or interfere with)	Atropine	Extreme dilation (indirect action) [dry mouth, fuzzy vision, difficult urination]

Look at all the symptoms of stimulation, the fight half of the fight-or-flight argument. Your eyes are dilated, you're running through the forest, you're being chased by something that can run almost as fast as you, and it's hungry. You want to get through that forest. You want to be able to use all the light you can get. Your eyes are dilated in a dark, moonless night to make the best chance of not missing trees that you can go up, and hopefully that thing behind you can't climb. So you want all the light you can get. You lose your bladder control. What is the matter? Urine is running down your pants leg but it's going to eat you anyway. [Laughter from class.] That's minor. On you go. [More laughter.] You're completely wound up. You are charged, the adrenaline is coursing through you, you can lift trees that are in your way, so to speak. That birth of energy is there. Believe me, you're not going to stop and enjoy a nice meal along the way. [More laughter.] Your appetite has disappeared. That's no longer a part of your process. All this is a consequence of a sympathomimetic drug, adrenaline, a drug that acts upon the sympathetic nervous system. And indeed, your eyes dilate because you have only a certain amount of innervation going to the sphincter muscles, but your radial muscles have been dictated to be active and you get this expansion of the pupil.

[Directed to student] Yes.

STUDENT: Sphincter, can you spell that?

SASHA: Oh gosh! It's the muscle that is circular and tends to contract.

Several students overlapping: S-p-h-i-n-c-t-e-r.

SASHA: Sphincter. The most common one is the anal sphincter where the muscles will actually close off the anus, and it has to be worked against. So that is probably the most easily understood sphincter muscle. The muscles of the iris are circular and on their contraction tend to close down the iris.

[Directed to student] Yes.

STUDENT: Why would certain drugs, let's take one of the opiates, say, constrict and with amphetamines they dilate?

SASHA: You should be able to answer this pretty much. The ones that constrict will be parasympathomimetic drugs. They will act upon the parasympathetic system, which will act on the sphincter muscles, and tend to constrict that group.

STUDENT: But, they have radial linked up with the sympathetic system.

SASHA: I was talking about the parasympathetic, para-. Radial is the sympathetic. The sphincter is parasympathetic. So those are the two in opposition that keep in balance.

[Directed to student] Yes.

STUDENT: Is there a difference between sympathomimetic and para-sympatholytic?

SASHA: Ah, they both work in the same direction, but for different reasons. That's a neat question. Thank you for having read that little note and asking about exactly this point. Let's take a drug such as amphetamine. You get the eye dilation because you're activating the radial muscles that pull the eye apart. If, on the other hand, instead of having a drug that acts constructively on the sympathetic system, you took a drug that acted destructively on the parasympathetic system as a parasympatholytic, you're disrupting the opposite side. Instead of pulling the eye apart by activating the radials, if you take the parasympathetic, that is the sphincter muscles, and interfere with them, the eye dilates. Same result. Instead of this tension changing and becoming more intense, this tension stays, but the sphincter muscles are no longer activated and so they don't give as much battle in the other direction. You move in the same direction whether you stimulate the sympathetic or you interfere with the parasympathetic. The net results are very often the same.

Now, for example, we mentioned amphetamine being a sympathomimetic. We mentioned belladonna, atropine, scopolamine, they are parasympatholytic instead of being this [demonstrating on board]. The net result is that the eyes dilate. If you can go into an emergency room, there's a very, very fancy trick to tell them apart. A person comes in stoned, not in much coordination, not doing a very good job. You notice, "God! Look at the dilation of those pupils." So you go to the person, open the eye and put a flashlight across the eye. And what you're doing is giving a sudden insult to the eye, the eye is getting too much light. The normal response is to contract. And if you have stimulation, the eye dilation due to sympathomimetic actions, the sphincter muscles that tend to contract the eye are intact. They're not being interfered with merely because you were stimulating the radial muscles to tear the eye open. You're not interfering with the muscles that tend to keep it closed. And they're intact, so you have what's called reflexive mydriasis. Mydriasis is the expansion of the eye. You put a light across there, the pupil will go "whoosh." As the light is there, the pupil contracts, then it goes out again, because the sphincters are intact. They're not being interfered with at all. And so, they are working and they'll overpower for a brief moment to protect the eye from light by their normal parasympathetic action.

You take a person who is stoned on atropine or belladonna or sco-polamine or jimson weed, wandering around, not making much sense, with big, black eyes. We have a dozen names of plants that grow in every Safeway parking lot that will really cause your eyes to dilate. This tea they use, Asthmador, these teas that are sympatholytic teas let the eye dilate by inactivating the sphincter muscles; and by default, if you shine a light across that person's eyes, there's no reflex, no response to the light. Because the things that reflex are inactivated; the sphincter muscles are not working. And so, you can immediately say, "This person had reflexive mydriasis—probably speed or a stimulant or something," or "This person had non-reflexive, or un-reflexive or areflexive, mydriasis, probably into a parasympatholytic drug. Maybe he has taken too much atropine com-pound." This light reflex is an extremely valuable differential diagnostic tool in the emergency room. A sweep of a flashlight across the dilated pupils can immediately distinguish between the two separate classes of poisons, and can suggest appropriate intervention measures.

Jimson weed is our favorite one around here. It has big white flowers and grows as a weed. Thorn apple. I can't remember the other names. These are the *Datura* species, so another common name for that type of plant is datura.

So just by looking at a visible portion of the autonomic nervous sys-tem, you can get a general class of things to go after. And, of course, if you are treating a person who's in a sympathomimetic situation, what you probably want to do is introduce a sympatholytic to quiet down that stimulation. A person who's in a parasympatholytic, you may want to apply a parasympathomimetic. So the treatment would be different depending on what type of drug has been used or what type of problem. These things can come from spontaneous situations in the body itself, not necessarily a drug effect.

[Directed to student] Yes.

STUDENT: What sort of drug would be considered sympatholytic?

SASHA: Oh, you may have used such things as some of the major tranquilizers. Haldol is often used. Partly because it has that direction of drug effect and partly it tends to modify the displayed symptoms of excitement of the sympathetic nervous system. Haldol would be an example.

STUDENT: They use Haldol to bring down people on coke. Cocaine.

SASHA: Cocaine is a sympathomimetic. You would not use Haldol for a person who has overdosed on *Datura* because you would probably make the situation worse. So yes. But cocaine is a distinct stimulant.

So this is kind of the feeling of the nervous system. Notice there's no arrow going straight out to the outside world [referring to the drawing on the board]. There is no way (outside of intense belief, meditation, or religion) of getting a direct effect at a distance from you, by means of your nervous system. If you want to make a dent in the wall, you pick up something and throw it at the wall. There is no tele-efferent counterpart to the tele-afferent input from a distance.

But generally, again, the feeling of things going in is afferent. Things going out are efferent. And going out, they go out to the voluntary or the involuntary system. And the involuntary, the autonomic system, is in these two branches, the sympathetic and the parasympathetic, that are all-the-time keeping you where you are. If it shifts to a place over here, either this is a stronger pull, or this objection to the pull is less intense. Either one will move one way or the other by either a stimulation of something, or by the loss of stimulation of something that is in counteraction to it.

So in essence, you have this whole body of nerves in the body. In the handout, I want certain terms to be familiar. The chemical structures are really quite incidental. Let me take a more general picture of the nerve system. You have a nerve, this is, in a very crude sense, coming along here [gesturing] which comes close to another nerve that goes on. This is the direction of the nerve flow.

Nerve conduction—the synapse. Let me present a somewhat simplified picture of the process that explains how a signal goes along a neuron, and especially how it gets from one neuron to another. On the nerve itself, let's pretend this is part of another big nerve over here, if you were to insult this end of the nerve [indicating], there is an upset of electrical charge. The nerve will then conduct a signal. The transmission of the signal along a neuron has been likened to the conduction of electricity along a wire, but nerves are not like electrical wires and that is a terrible analogy. Actually it is misleading in that an electrical signal moves at the speed of, well, electricity, and the wires involved must actually touch one another to transfer the signal from one to the other. On the contrary, in the neural network the signal moves much slower, at about twenty meters per second rather than at 300 million meters per second, and the system won't work at all if there is actual contact between sequential neurons.

How does a neuron conduct? The body fluids are filled with salts of various sorts that maintain a needed ionic pressure everywhere. But there is an active, energy-consuming process continually taking place that keeps the insides of a neuron exceptionally rich in potassium ions, as opposed to the fluid outside the neuron which, as with all other body fluids, is richest in sodium ions. You have a nerve sheath, a nerve cell membrane, which has a charge across it due to selectively moving calcium ions out and potassium ions in. This unusual distribution keeps a negative potential (about negative seventy millivolts) across the nerve membrane, and this is the loaded mouse trap that allows conduction. If you were to stimulate a nerve somewhere in the middle of it, say by sticking a pin into it, this delicate balance would be upset, the membrane would lose (for a short moment) its ability to keep all those sodium ions out, and when they flood into the neuron, the potential disappears. What you're doing is disrupting that charge by permitting an influx of calcium ions, which then further propagates that disruption of charge while simultaneously "healing up behind itself" due to the movement of potassium and sodium ions reestablishing that original polarity. Like a pebble tossed into the middle of a quiet pond, this disturbance, in turn, disturbs the areas around, and this loss of voltage (depolarization) spreads outwards in all directions. No, this is not a good illustration, because the ripple suggests multiple waves and a local oscillation. The action of nerve firing is a one-time event occurring here and moving everywhere else. Better to visualize the pinprick as a match tossed onto the middle of a flat surface that has been covered with a thin layer of gunpowder. Ignition occurs at the point of contact, then spreads in an ever-increasing circle leaving behind it exhausted residues that cannot burn. Since the neuron is a long, thin tube rather than a flat surface, the disturbance would quickly meet itself on the opposite side of the tube, and then spread (at twenty meters per second) towards both ends of the nerve. And as that signal spreads away from the site of insult, the membrane behind this moving signal is exhausted (refractory) for a short time, but it heals up quite quickly (repolarizes) and becomes responsive again. There are pumps continually working to reestablish the excess of potassium over sodium inside the neuron, thus reestablishing that original negative seventy millivolts, the resting potential. And as it disrupts the charge, it tends to move that disruption from one end of the nerve to the other. If you disrupt the nerve by putting a pin in there, you'd have a signal that would suddenly have a disruption at this end a few milliseconds later. If you took a nerve, and got in there and stuck a pin in this end, it's hard for the signal to go the other way. We get it at this end, nothing would happen because there's no way of getting across the gap to the next nerve. The direction of nerve impulse is based on the fact there are neurotransmitters located here. But nerves aren't normally stuck with a pin in the middle, they are stimulated at a receptor on one end by a thing called a neurotransmitter (all of this takes place in this little area, the synaptic cleft), and diffuse their way across to the other nerve and insult it, and start a nerve signal going out again.

The synapse is a narrow gap that occurs between two neurons, and it is named according to the neurotransmitter that is needed to complete the transmission of a signal across it. It is that doing-of-things to this transmitter (releasing it, prolonging its life, destroying it, imitating it, keeping it from being made in the first place) that explains the action of most drugs that affect the nervous system.

The figure in the hand-out is a rough schematic of a synapse.[2] This is called an adrenergic synapse, because the neurotransmitter involved is adrenaline (the British name for the chemical), which is called epinephrine in the United States. Actually, the neurotransmitter involved is norepinephrine (which would be called noradrenaline in England). So to be exact, the synapse should be called the noradrenergic synapse, but it usually isn't. The same general scheme would apply to a cholinergic, or dopaminergic, or serotonergic synapse except each of these would employ either acetylcholine, or dopamine, or serotonin as the transmitter, instead of norepinephrine.

In this figure, the nerve coming in from the left is the afferent nerve (reasonable term, as the signal is arriving from there) and the area where it swells up just before the synaptic cleft (the gap between the two neurons) is called the presynaptic nerve ending. The cleft itself is maybe 200 angstroms wide, which is the length of a few molecules end to end. This is the space that the released neurotransmitter has to diffuse to affect the transmission of an impulse. And, quite logically, the area on the right of the cleft is the postsynaptic nerve ending, and the nerve itself is the efferent nerve (from the point of view of the synapse). The structures that release the neurotransmitter (the neurotransmitter storage sites) lie only on the presynaptic side, and the structures that respond to the neurotransmitters (the neurotransmitter receptors) lie only on the postsynaptic side. Thus, the transmission across the synapse can occur only in one direction, from pre to post.

So the direction of the nerve's operation is a function of the fact that neurotransmitters are here, and the receptors are on the opposite side. It takes diffusion across that cleft of a neurotransmitter to keep a nerve signal going on. So you have a relatively, compared to electricity, slow motion movement down a nerve. The transmission of a signal is much,

---

2 Adrenergic Synapse handout on page 163

much slower than the speed of electricity because of the ionic diffusion in and out of the neural membrane (the depolarization that allows impulse propagation along the original neuron), and also because you require the actual chemical diffusion of the released neurotransmitter across the synaptic cleft to the other side (the receptor activation that allows impulse initiation in the new neuron). The distance is close, but they do not touch. They do in certain lower animals, in very simple neuron systems, largely in simple marine organisms, these actually touch and there's no need of a neurotransmitter for those systems. In human beings and higher animals, you do have that opening and you require a neurotransmitter.

If you were to stick a needle in the center of the nerve, the signal would go in both directions. One would not achieve anything when it got to the end of the nerve, but this one [indicating] would continue the process going. So all of your afferent nerves, all the nerves that come in, people do not realize that these are really kind of hairy, long things. Let's say you take a pin and you touch the pin to the tip of your finger and somewhere upstairs says, "Pain!" or "Ouch!" You are actually insulting a nerve way at that end of the nerve down here. Where is the other end of that nerve? That is now starting a process of depolarization and polarization and the signal is going along the nerve. People say it goes to the first synapse. But where is that first synapse? The first and only synapse of that nerve that is sensitive to pain is in the brain.

The picture is totally cloudy, since it is virtually impossible to study any particular aspect of it in isolation. Everything interacts with everything, and the simple measurements of transmitter levels following the administration of this, or the sectioning of that, rarely provide any consistent explanation of higher neuron function. One of the complications in the study of afferent signals is that the nerves involved cannot be gotten at in isolation without doing damage to them. The signals that pass from the brain to the muscle or the gut, the efferent neurons, are several in number, and are connected in series through accessible synaptic junctions. But the nerve that starts at your fingertip, and which, when insulted, registers at pain or heat or touch, passes up the arm, across to the spinal tracts, up to the brain, and the other end of it is lost some-where up there in a maze of "green and red wires." The touch neuron does not even have a known transmitter. Probably it is associated with a seemingly chaotic collection of many neurons and neurotransmitters. Not chaotic, but truly beautifully organized. It merely seems chaotic once inside the brain. That nerve, if you were to take it and follow it along with a pointer, or haul it out and look at it as it comes out, goes all the way in here, all the way into the spinal column, goes up, crosses over, goes up the other side of the spinal column, and up into an area in the central part of the cortex that is known as the somatic homunculus.

An aside, with the word *"homunculus."* This is literally a little man, a manikin (Dutch, little man), which was transformed in French to man-nequin, to mean a life-sized model or dummy. But the Latin original, homunculus, has a colorful history in the world of medicine and anatomy. In the marvelous era after the function of sperm and egg were known, but before the atomic theory put to rest forever the concept of infinitely dividable and infinitely small, it was believed that the entire new indi-vidual was contained preformed, but just very, very small, in the head of the sperm. And, of course, within him was his progeny, and within the progeny, the progeny's progeny, and so on for all future generations. All pre-cast. Homunculus within homunculus within homunculus. The period between the acceptance of the concept of the infinite and indi-visible atom (speculated upon by the ancients but not generally believed until the beginnings of the nineteenth century), and the discovery of the chromosome (about 1870), represented one of disordered confusion in the area of reproduction. Even the chromosome made little sense until the role of DNA became evident, only within the last few decades.

The image of the homunculus still lives in the portrayal of the sensory and the motor cortexes. A homunculus means a "bitty man," and this is a bunch of tissue of which this is the finger and that's the toe and this is the tongue and this is the nose and that's the genitals and this is the elbow, the "other ends" of the sensory neurons from here and there all over the body can be used to draw a small, misshapen human in the sensory region of the cortex. And there is a second, a motor homunculus, lying close by, to connect, nerve for nerve, out again to the real owner's outer body, outlining marvelous figures, with immense thumb and foot bottoms and genitalia, but little emphasis on the small of the back!

That very magical little misshapen gnome is the area where that neuron, which starts at the finger, goes to. There is nothing in between. These nerves are very little understood, because you can't get at an intermediate spot and study both ways. Nerves coming out of the body, no problem. They go to synapse after synapse. Their signal eventually reaches the brain, there's a synapse along the spine, and there's a synapse outside the muscle endplate. These are all different areas you can study the nerves going out. But going in, that is harder to do because they're lost in the mush of the brain.

What are the neurotransmitters that are responsible for conducting a pain signal from the tip of the finger to the brain? I don't know. It's not really known because you can't isolate that end of it. The neurons that touch neurons in the brain are a horrible gemisch. Basic neurotransmitters in the periphery have been very well studied. I'll give you the terms now and then we'll get back into a bit of it in the next hour. You should know norepinephrine is often called NE. Epinephrine, is often called "epi." Epinephrine is adrenaline. Norepinephrine is noradrenaline. Noradrenaline, adrenaline are largely British English. Epinephrine and norepinephrine are largely American English. Same chemicals. You must know, eventually, one way or the other, you're going to come across it enough times, serotonin, often relayed as 5-HT or 5-hydroxytryptamine. And one more I think we'll add to this list, is dopamine, often abbreviated DA. And one more, acetylcholine.

In the peripheral, the autonomic nervous system, acetylcholine and norepinephrine are the major players. In the brain these both play a role, and probably serotonin and dopamine are amongst the primary ones. This is in the CNS. All of these are mandatory to get a signal from one nerve to another, and the communication between nerves requires these types of neurotransmitters.

Neurotransmitters are the chemicals that are fed into the synapse, wander around and get to the other side of the synapse. There is one and only one function for a neurotransmitter. It is to bridge, briefly, that narrow gap between the pre side and the post-side of the synapse, and thus to allow the propagation of the nerve impulse to continue its course. The action of many of the nerves affecting the nervous system can be explained by their relationship to the single function. The actions of many drugs are closely and intimately tied up with these neurotransmitters.

A corollary to the presence of a neurotransmitter for successful transmission between neurons is the need of destroying the transmitter once it has done its job. As mentioned earlier, neurons need time to reestablish the resting potential across their membrane, so as to allow additional signals to be transmitted. This refractory period is a real rest time that is mandatory. If the neurotransmitter stays in the cleft pushing against receptors, always trying to re-trigger a post-synaptic signal, the neuron can never completely recover. There are neurotransmitter destroyers both in the cleft (to sweep the area clean readying the system for some future impulse) and in the presynaptic area (to inactivate any transmitters that may have been released but hadn't yet made it to the synapse). Clearly if these destroyers are themselves interfered with, the neurotransmission becomes more frequent and stronger. Thus, yet another mechanism of drug action.

Let us look at the adrenergic synapse system in the figure from two points of view. One, the route by which the neurotransmitter itself (norepinephrine) gets there, gets used, and gets destroyed (normal function). And then how drugs can do something at almost each stage of this normal process (drug-modified function). Here the normal process will be followed, and in subsequent lectures, the various drugs that excite, depress, or otherwise influence this picture can be located against this roadmap.

The neurotransmitter for the adrenergic system is norepinephrine, and it is synthesized right in the presynaptic nerve ending itself. The process leading up to it starts outside the neuron, and in the case of neurons in the brain, outside of the brain. A fundamental amino acid, tyrosine, is brought from the bloodstream into the brain (through the blood-brain barrier by an active transporting mechanism) and it is also transported directly into the neuron to serve as the starting stuff for eventual norepinephrine. The "dirty pictures" for this series of chemical manipulations are drawn out in Figure 2. The symbolic location of these intermediates are drawn in Figure 1. Tyrosine can be generated in the body from a yet simpler amino acid, phenylalanine. Both of these

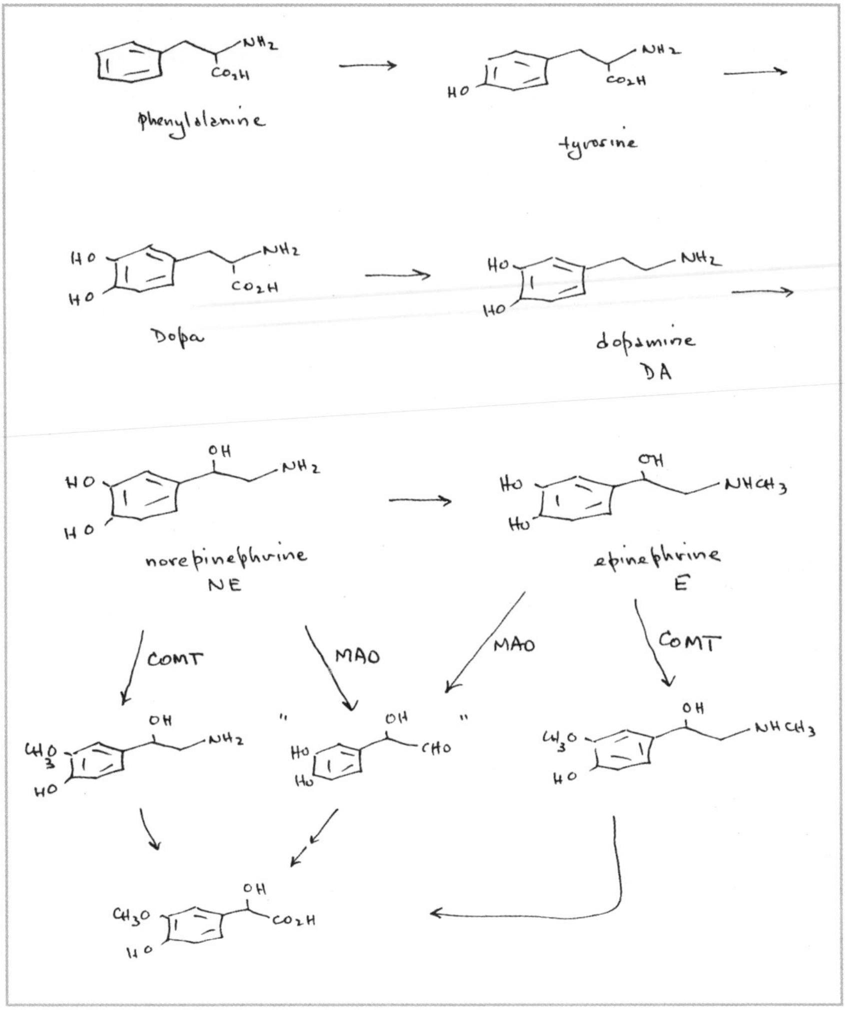

Figure 5.3: Dopamine, norepinephrine, and beyond

amino acids are derived from proteins in food. Tyrosine is mandatory for eventual proper nerve chemistry and function.

Tyrosine is hydroxylated to a catechol amino acid called DOPA (from the chemical name, dihydroxyphenylalanine). This amino acid must be created within the cell (and in central neurons within the brain, of course) as only a modest amount of it is transportable across the blood-brain barrier. It, in turn, is decarboxylated to provide dopamine (DA) which is a major neurotransmitter in its own right, in the brain. Dopamine can eventually be hydroxylated in the beta-position (DBH, dopamine beta-hydroxylase) to form norepinephrine (NE). This is one of two ultimate neurotransmitters in the peripheral nervous system. The other is acetylcholine (ACh). This transmitter is stored in small vesicles located throughout the presynaptic nerve ending near to the edge of the synaptic cleft. When a nerve impulse arrives, some of the NE vesicles release norepinephrine, and some of this escapes into the synaptic cleft. Meanwhile there are metabolic scavengers at work, destroying all nor-epinephrine both inside the cleft and inside the neuron. In the synaptic cleft, NE is removed largely by methylation of the catechol portion of the molecule by the enzyme catechol-O-methyl transferase, COMT, but some NE is actively reabsorbed by the presynaptic membrane. Within the presynaptic nerve ending, NE is removed largely by oxidative loss of the amine function of the molecule by the enzyme monoamine oxidase, MAO. Also within the neuron, some of the NE can be regathered back into a NE vesicle by an active uptake mechanism.

Every stage of this scenic tour can be a point of action of drugs that affect the adrenergic neuron synapse. Stimulants in general increase synaptic transmission; by encouraging the release of norepinephrine (indirect acting agents such as amphetamine and mechanical means such as electroconvulsive therapy), by it entering the cleft and fitting directly into the post-synaptic receptors (direct acting agent such as epinephrine), or by inhibiting the oxidative destruction of norepineph-rine (monoamine oxidase inhibitors such as tranylcypromine), or by inhibiting the reuptake of norepinephrine from the synaptic cleft itself (reuptake inhibition by stimulants such as cocaine and antidepressants such as imipramine).

It is easy to imagine the counter-mechanisms that can be found to operate in the actions of some depressants, actions that effectively decrease the amount of norepinephrine available for action inside the synapse, whether it is through the release of some imitation of norepinephrine that is not very efficient (false transmitters such as alpha-methylnorepinephrine for treatment of high blood pressure) or the blocking of epinephrine re-uptake into the storage vesicle (as with reserpine used towards the same goal).

The picture of the noradrenergic synapse is the best understood, but presumably all neural systems work in some similar manner.

The companion of norepinephrine in the autonomic nervous system is acetylcholine (ACh). The synapse that has it as a neurotransmitter is known as cholinergic synapse. It, too, is a narrow separation between a pre-neuron and a post-neuron, generally a little wider than the adrenergic synapse. Similarly, upon presynaptic stimulation, acetylcholine is released into this cleft, and in diffusing across to the post-synaptic membrane, activates specific receptor cells effecting depolarization and thus conduction of the signal by the post-synaptic neuron.

The problem of neurotransmitter removal is exactly the same as seen with the adrenergic system. If the acetylcholine remains too long in the cleft, the system activated by the postsynaptic neuron stays continually activated, and fatigues to eventually become unresponsive. This condition is normally controlled by an enzyme system that specifically destroys acetylcholine. And as ACh is an ester, this destroying mechanism is an ester hydrolyser, or an esterase. The natural enzyme of deactivation is acetylcholine esterase, AChE. Many of the poisons that we will be discussing later, things such as insecticides and nerve gases, are inhibitors of this enzyme system, and are called acetylcholineesterase inhibitors (AChEI). Since they permit the parasympathetic system to be continually activated (muscular tetany, intestinal cramping, hypotension) it is completely logical that treatment of such poisoning would be with agents that block (interfere with, "lyse") the parasympathetic nervous system (parasympatholytics such as Atropine). The cycle of the synthesis of acetylcholine is completed (see Figure 3) by the reacetylation of choline with acetic acid.

As with the adrenergic system, there are drugs that can interact, agonistically or antagonistically, with all aspects of the cholinergic synaptic transmission system. The interference with the removal of the neurotransmitter by esterase inhibitors has been mentioned. This can be in a reversible manner such as the carbamate insecticides like Sevin and Zectran, wherein normal function is recovered rather quickly, or irreversibly (such as with the phosphate insecticides such as malathion and parathion, and neural war gas poisons such as tabun and sarin) where the enzyme is permanently destroyed. There are drugs (such as

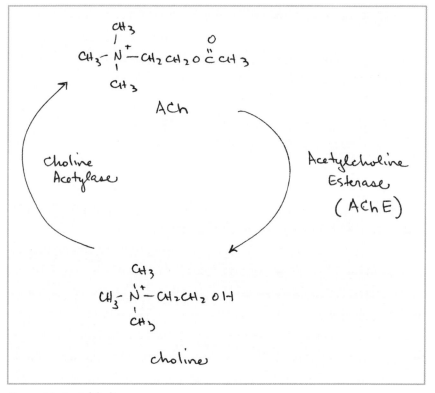

Figure 5.4: Acetylcholine

pilocarpine, used in glaucoma for the dilation of the canal that drains the aqueous fluid from the anterior chamber of the eye) that are directly acting upon the post-synaptic cholinergic receptors. There are drugs that can interfere with the reacetylation of choline, thus robbing the synapse of its needed transmitter (such as the botulinum toxins).

The dopaminergic system is again similar in general structure to the adrenergic, except that the neurotransmitter is dopamine rather than norepinephrine (see Figure 2). The dopamine system is exclusively within the central nervous system, as compared with the adrenergic and cholinergic, which occur on both sides of the blood-brain barrier. It is probable that dopamine neurons do not occur in isolation, but rather they intimately interact with, and exist as extensions of, other neurotransmitter systems such as acetylcholine, serotonin, GABA (gamma-amino butyric acid), and certainly some of the peptide neuron synapses. Dopamine and acetylcholine have been argued as acting in some sort of harmony in the explanation of the drug-induced symptoms associated with the use of tranquilizers in the treatment of psychosis.

Serotonin (or 5-hydroxytryptamine, or 5-HT) is the only indole neurotransmitter that is established with certainty. It is the second major transmitter (along with the dopamine) that functions broadly within the central nervous system. Its structure, and its relationship to the precursor amino acid tryptophan, and its potential derivatives that are centrally active hallucinogens, are given in Figure 4. It may be the major factor in the calming effects of sedation, and may actively promote sleep.

The GABA-mediated neurons are widely distributed throughout the brain, and are thought to be generally inhibitory to the action of other neurons.

This has been a rather deep dipping into the wiring of the body, which is the background of the actual functioning of drugs, what they do at the molecular level, and allowing some classificational connection between drug and neurotransmitter. The next lecture will observe the action of drugs on more of an organizational level. What happens with the macroscopic system; what is actually observed when a drug acts on the body.

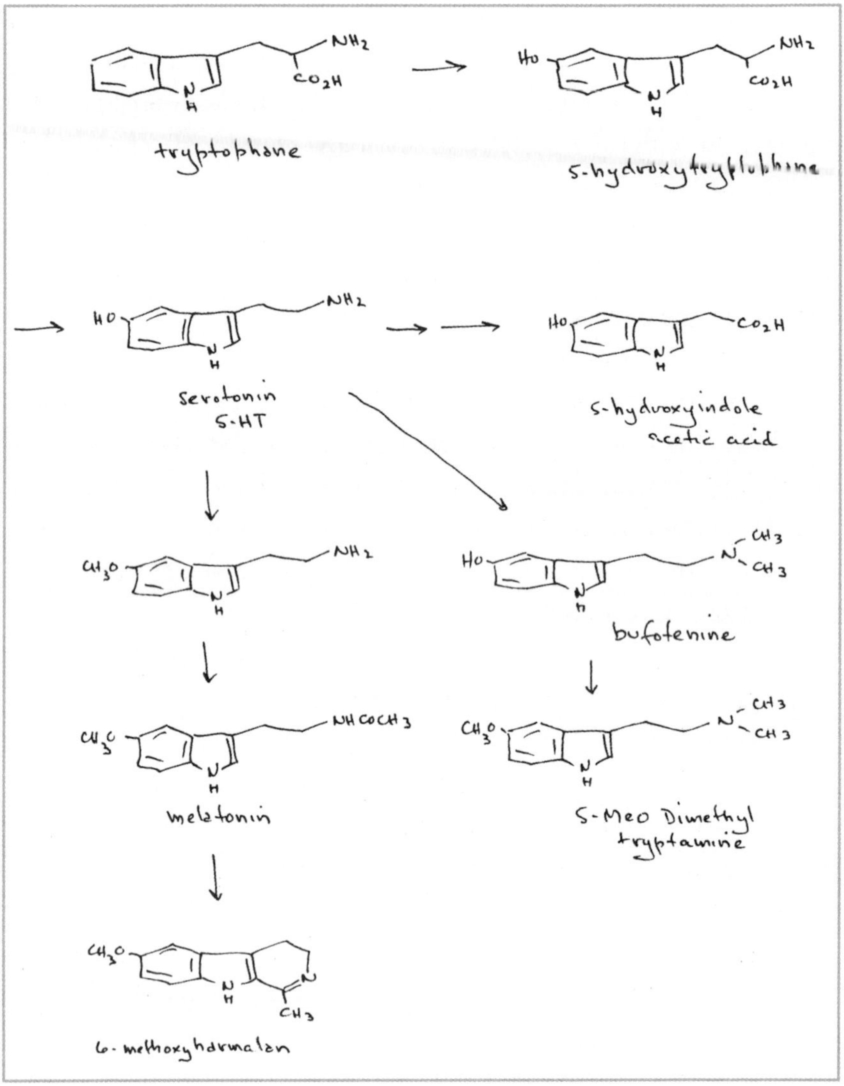

Figure 5.5: Serotonin

Okay, we're at the hour. I will go on with some of this in the next hour and try to tidy up.

Handout chem structures 1

Handout chem structures 2

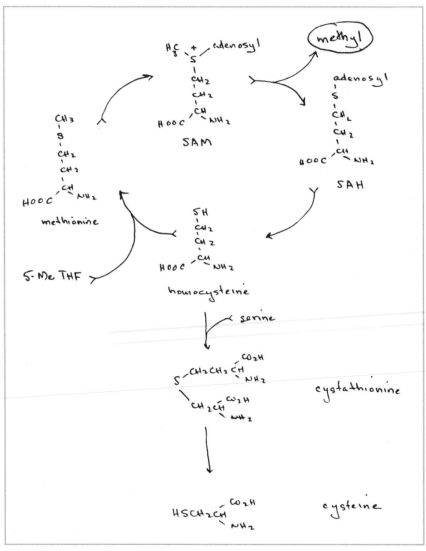

Handout chem structures 3

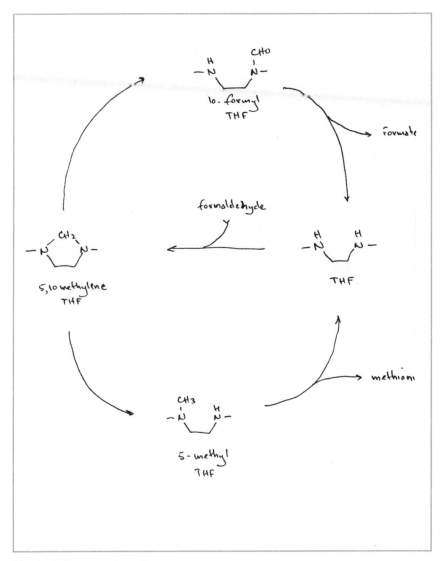

Handout chem structures 4

February 17, 1987

Drug Action

SASHA: Okay. Two-thirds of the people are here. Might as well get launched. The handouts—what I did was I began listening to some of the tape of the first lecture, and I realized that a lot of the definitions I had given probably too quickly for people to both take notes and listen. This is a problem I know very well, having been at that side of the desk for a long time. When you're listening to a person, you like to listen. And, when you feel you should take notes, you take notes. But somehow, when you take notes you don't listen. I mean, there's no way you can hear while you're taking notes.

So what I'm going to try to do is where I have definitions, I've written them out. There are twelve pages of them over here. In essence, these are the definitions of that first lecture. So they're there. Toward the end of this lecture, I want to get a little bit more into the flow of where we were going, which, once we got through the definitions, went into how drugs came to be in the body, how they get around in the body, and what happens to them to some extent. We didn't get much of that, because a lot of that is chemistry and I don't want to get into it. What happens to them in the body, where the nerves go that make things happen that the drugs inspire in the body. The framework was laid for the explanation of the action of drugs at the molecular level. The drug versus neurotransmitter relationship. That approach was completely logical and totally analytical, in that it was based on the fact that all affective states, from emotion to stimulation to depression to curiosity, are somehow functionally explainable (if not today, at least someday) by the tracing of the neuronal pathways, and all these pathways are glued together by the magic of neurotransmitters. It becomes very appealing to trust the chemical relationship between drug and neurotransmitter to eventually explain the effects of the drug. Today I want to get a little bit more into what the drugs actually do when they get the nerves inspired to do things, when they get to where the nerves are inside the body. Namely, to sort of continue this progression.

In real life, the drug affects the entire organism, which is a bigger operation than just a bunch of neurons and some fleeting chemicals that allow them to communicate. I would like to present an entirely different type of classification that comes closer to treating the individual as a single, whole unit, rather than a collection of a hundred billion interconnected neurons.

In a general sense, all drugs can be looked upon as being organizing or disorganizing, and they can be looked upon as being directed towards ill people or towards well people. This bizarre structure gives rise to the grid pattern shown below where I divided drugs up into very odd classifications. Drugs can be given to people who are ill to, in essence, keep them ill. Or drugs can be given to ill people to make them well. Drugs can be given to people who are well to make them ill. And people can give drugs to people who are well to keep them well. You can make a grid out of this.

Everyone has this first impression: Drugs are given to sick people to make them well, not to well people to make them sick. Well, if that were so, we probably would not have this class, because one of our main topics will be the administration of drugs to otherwise healthy people to intentionally induce a change of state which would, from the viewpoint of many observers, be classified as something quite abnormal and maybe even something sick!

If you have a person who is sick, there are only two things you can do. You can give them something to maintain that sickness, and this sounds a little bizarre. Why wouldn't you get them well? Well, maybe you can't cure what's wrong with the person. Let's say the person is missing an organ and because the organ is gone, something's missing in the body. Take the thyroid. We're going to talk later on about radioactivity. You

HEALTH STATE OF AN
INDIVIDUAL

	ILL (a patient)	WELL (a subject)
The drug is supportive (organizing)	SUPPRESS SYMPTOMS	MAINTAIN STATE
The drug is disruptive (disorganizing)	ELIMINATE SYMPTOMS	GENERATE SYMPTOMS

As examples:

Suppressing symptoms. This represents a cosmetic approach
to an illness. You supply the individual with what is needed to
overcome the problems of illness, but there is no move towards
repair. A person has diabetes mellitus because his pancreas is
inadequate, so he takes insulin. This does not help the
pancreas.

Eliminating symptoms. This is the direction of cure, and
there must be disruption by a drug to achieve this. It must kill
a bacterium, or irreversibly bind to a poison, or destroy the
reproductive capacity of a cancerous cell.

Maintaining the healthy state. Drugs in this category are
primarily for the prevention of disease. Innoculations,
antibodies against tetanus and polio, the pre-treating of the
body with eventually needed defences. And if pregnancy can be
considered a temporary pathology (the rapid growth internally, of
a body that is not of your unique chromosome make-up, and
inevitably fatal if not rejected at some critical point) then
contraceptives fit into this category.

Generating symptoms. Certainly, there is a large body of
drugs with the purpose to make a well person "sick." In the more
accepted areas, there are drugs such as the anesthetics used in
surgery -- from an awake, conscious person one gets a comatose,
unresponsive person. And in the less kosher areas, the use of
consciousness-altering drugs should qualify for this category.

1

can give a little bit of radioactivity and make the thyroid visible by the radioactivity in the form of iodine that goes into the thyroid like water into a sponge. But you can give a big pile of radioactivity and cook the thyroid. If you give enough radioactivity, say give 150 or 200 millicuries (mCi) of iodine-131, it'll go into the thyroid and the thyroid will be radiated and destroyed and you end up with no thyroid.

Why in the world would you want to do that unless you have a thyroid that's cancerous or a thyroid that is totally out of control? What you need the thyroid for is for certain thyroid hormones. But if you, for some reason, had to get rid of the thyroid because of cancer, because of some mismanagement, because of some rejection—some people don't like their thyroid—there is a disease known as Hashimoto's disease, which is autoimmune rejection. It's an immune response. We'll talk more about the immune response when we get into radioimmunity, immune radioactive assay, and EMIT urine screenings. But there is this immunological response in the body in which something foreign gets in the body, the body rallies all these marvelous bits of chemistry and says, "Hey, you're foreign. I'm gonna gather you up and get rid of you or I'm gonna digest you, or I'm gonna somehow inactivate you from hurting the body." And sometimes the body gets this weird little thing in its head that some part of the body that's normally in the body is foreign. And this can happen with the thyroid. It's one of the more common spots. It's called Hashimoto's disease for the person who first described it and put its character together. The body says, "Hey, that's not my thyroid, that's some other person's thyroid" and it mobilizes the whole immune system and gets rid of the thyroid. And pretty soon you end up with no thyroid. And a person with no thyroid lacks all the hormones that come from the thyroid.

And so, here you have a person who's sick, but you don't give them a drug to give them a new thyroid because that just doesn't work. Someday you could write a DNA code and do something about it, but you can't now. So what you do is give them the hormones that the thyroid would provide. And then a person can function as if they had a thyroid because they have been given these drugs. So what is being done is giving the person a cosmetic.

It's very much like the use of insulin in diabetes. How many people are familiar with the term "diabetes" in its general sense? Most. There are two types of diabetes: diabetes mellitus and diabetes insipidus. I'll write these notes out and you can have them some day.

Just the flavor of it, diabetes, the really serious one, is diabetes mellitus. It is the one in which there is an inadequacy of the pancreas to produce insulin. Insulin is a very necessary polypeptide—I guess it's a drug because it does something, but it's generated in the body so we don't think of it as a drug—that causes the cell to be able, in the area when glucose is available, to take in the glucose and use glucose. So insulin is a hormone because it's made here and it's used there. The usual sense of a hormone is where something that regulates body behavior is manufactured in one place in the body and is used somewhere else. So in this case, the hormone from the pancreas is called insulin. It's distributed in the blood. There is sugar in the blood. Blood doesn't need sugar, it's the cells, the tissue, the thing that blood gets to that needs the sugar. And sugar can be right up against the tissue and it won't go in, it won't be utilized unless there's insulin. Insulin is the thing that makes the cell accept the sugar and use it.

[Directed to student] Yes.

STUDENT: I thought insulin was to break down excess glucose.

SASHA: Glucose breakdown—I don't know. If that is a role of insulin, I don't know of that as a primary role.

STUDENT: It has also a secondary role.

SASHA: I'm not familiar with that. I know the expression that was used is: Cells swim in a sea of glucose that they can't use in the absence of insulin. So it's utilization. Utilization of glucose involves its destruction.

But certainly, the use is dependent upon insulin. If you don't have a pancreas or the aspect of the pancreas that creates insulin is wiped out, you have no choice but to give the person insulin. Well, this is why until recently, people who like these little sweetbreads (the thymus, the pancreas are called sweetbreads) never get pork sweetbreads. Because the pancreas of the pork is so close to human insulin that the pork pancreas has been used to get the insulin out to provide insulin to people who do not have their own pancreas and need insulin shots for diabetes. Now, it's not quite the same and one of the very, very fortunate things for millions of people without insulin, up until quite recently, has been that the porcine insulin is so close to human insulin, one amino acid here, maybe one amino acid down there different, that the human immunological system does not see it as being different. It says, "Okay, there's the insulin. I'll function with normal insulin." Actually, it's a little bit different. And in time you can build up an antibody to it because the antibody system is geared to recognize things that are not part of the normal body that produces the antibody. Porcine insulin is different, but not much different. It's so close that it could be used in many cases for the bulk of a person's life without much increase in dosage because the antibodies were not triggered by that foreign protein.

Now they make insulin by taking the true genetic code for human insulin, implanting it into bacteria. The human insulin when planted into bacteria, the bacteria reproduce bacteria, as bacteria are prone to do, but in the course of that, in duplicating, it duplicates this part of the DNA which says, "I'm human insulin" and that produces insulin. You somehow filter out the bacteria, get the insulin, and you have a drug now that is the perfect substitute for human insulin. It is human insulin.

[Directed to student] Yes.

STUDENT: Why do you have to inject insulin? Why can't you take it as a tablet?

SASHA: Okay, a good point and it has a very straightforward answer. Remember, in the cored apple, where you go in the mouth, down through the stomach and into the intestines; the stuff that goes through the intestine wall and gets into this portal circulation that goes over to the liver, very few things will go through that wall. It is a very permeable wall, but only to small things. Sugars will go through, ions will go through, water will go through, amino acids will go through. Polymeric sugars, such as cellulose, won't. Polymeric amino acids, such as protein, won't. Insulin is about a forty- or fifty- amino acid small protein; it won't go through. The only way it will get through is to be digested. The digestive process tears these big polymers down to their components. The components are absorbed and then the liver and the enzymes in the body resynthesize them into the proteins that are wanted.

So, if you take a protein, unless it has a specific poisonous effect, or a specific property that is bad, it is digested to its components, the components are absorbed, and the absorbed components then are reutilized by being repolymerized by the direction of the enzymes in the body. So it doesn't matter what protein you take as long as the amino acid is the amino acid you need. So, the idea, "This protein is especially good for me, that protein is not very good for me" does not have any validity as long as that protein contains the needed amino acids and is digested, can be digested, and the amino acids go through. So insulin going in by mouth will become proteins, it'll go in the body as proteins, there is no insulin.

STUDENT: So it's destroyed.

SASHA: It's destroyed. All proteins, all polymers—no, not all. With small children sometimes there is that capability; in newborns especially. Newborns are weird. [Laughter.] They have chemistry and biochemistry all of their own that very rapidly changes. Who has actually handled a child less than one or two days old? Look back to that time. You're handling an organism that is not quite yet put together in a human way. I mean, you have the fontanelles, all these sorts of things that repair over a few years, the holes in the head seal up. But if you take a baby—there's a beautiful neurological test that's known as the Babinski in neurology. It's where you take a person's flat of the foot and you go "rrrmmph" on the sole. Probably someone's done it to you. It's really "nnnyyaah!" You know, your toes immediately flare. But *your* toes flare up, that is the normal response when you go over the sole of the foot. There is what's called a Babinski when the toes curl down. It means something is screwed up with the nervous system. Newborn babies? The toes curl down. Three days old? They go up as they should. The nervous system is not quite put together.

What I came across just three weeks ago, I was amazed. Back to the half-life. You remember, the idea of the plot and how fast something goes out of the body? The half-life of caffeine in human beings is about three or four hours. The half-life of caffeine in newborn babies is about four days. Twenty times longer. The enzymes that are necessary to metabolize caffeine haven't been developed yet. You take a newborn baby, still wet and little bit discolored from whatever comes out with baby, toss the baby in a warm swimming pool. "They'll drown!" Nonsense! That baby will swim like an Olympic champion across the pool to the edge. [Laughter.] Three days later, the baby will drown. Take a newborn baby and hold your fingers out like that and let their still damp fist grab your index finger. That baby will grab, and hold on, with extraordinary strength! Those fists are glommed on to those fingers. Maybe from something when they were born in trees, you know, you're holding onto the limb of the tree while the wind blows, or to the mother's hair while she takes all the afterbirth away, or goes to find things. Fine! But during that first couple days that gripping thing is a death grip. Three days later, it's gone. All of these things are present only in the first days. The heart has just gone through a tremendous revolution because the heart wasn't pumping anything to the lungs. Now all of a sudden with the birth trauma, you open this valve, close that valve, flap this together, that blows open, this seals shut, and the whole circulation of the blood in the body is totally changed. And that is tremendous trauma of neurological readjustment the first couple of days. And prematures often take longer to do this. The whole treatment of prematures has often been considered a small-scale treatment of adults. The treatment of children: "It's a small child, we'll give them a small amount of drug." Totally unaware of the fact that drug could be handled totally differently. And it may be totally inappropriate. Or a larger amount may be needed because it doesn't get in.

The whole thing came up because most babies, as far as I know all babies, in the first day or two, their whole digestive tract is absolutely nuts. I mean, they're not used to taking milk while they're still in the uterus. And all of a sudden, you're pounding a whole new food through a new hole and the whole process of digestion has never really had a chance to find its footing. It's been getting all these lovely things from the umbilical cord. And so, the gut cannot reject proteins. So certain proteins that normally would be chewed up and digested go right into the child. And you can get toxic responses. People who have been involved with and around newborn babies know that a baby cannot be exposed to this particular protein source or that particular food source for quite a while because the body is not geared with the mechanisms for getting rid of these things.

Let's say a person is running high blood pressure and you want to do something to bring the blood pressure down. Why do you want to bring the blood pressure down? So you don't blow a pipe somewhere in the brain and have what's called a CVA; the common term is cerebral vascular accident. It's, in essence, an aneurysm, something in the brain, a vein that's gotten ballooned—you've seen this when you take a rubber tube and you put too much water in it and all of a sudden it blows up to a sausage. And, by golly, the water's going in, the tube's integrity is no longer there because it's suddenly been popped out like a balloon. You blow a balloon, you huff and puff and you can't get it to open, all of a sudden, something gives way a little bit and the balloon expands. Veins are the same way. They have an integrity of the structure that holds the blood. Arteries are the same way, too.

But if you get, for some reason, a weakened place, the vein could blow up like a sausage. You have a small vein and suddenly you have a large vein and then it goes down to a small vein again. It's called an aneurysm. And it is an extremely hazardous situation because once the vein has blown out that way, it doesn't have the intrinsic strength anymore, and any little push, any little additional pressure that normally would be resiliently maintained or controlled will pop it. And once you get bleeding inside the brain it's a bad situation. You have blood flowing in areas where there should be no blood.

I had a good friend who called me up from a hospital down near Redwood City. I hadn't seen him for several months and he said, "I wanted to let you know I'm in the hospital and you almost killed me!" I said, "What did I do?" He said, "Well, you encouraged me to lose weight." I said, rather humorously, "I'm sorry if that was a life-threatening situation." He said, "It *was* life threatening." He was 230, 240 pounds and I encouraged him to at least get that down to something that's reasonably overweight instead of excessively overweight. And he did. He got going down to 200 and he said, "Hey, this is feeling pretty good." Went down to about 180. Down to 175. And he felt really better than he had in years. He was about fifty-five years old.

And he was showing off his chest to his family and his friends, patting this now available part just under the ribcage, and he felt this strange lump down there. Never felt it before. It was all fat before and he couldn't feel a thing. Went to his doctor, the doctor saw it and felt it, and into emergency surgery. He had an aneurysm all through there. It was about the size of a large salami about seven or eight inches long and blown out like that. If that were to break, you have about a minute or two in surgery to save a life. Blood is just pouring out into the body, you've only got five or six liters of it, and when it's poured into the gut, you have a gut that's filled with five or six liters of blood and there's none left to circulate. You are dead.

So into surgery, no psychological preparation, right into surgery, opened him up, cut off the vein, removed the aneurysm, hooked the vein back together again. And he's recovering very nicely. He never would have known it if he had not lost the weight. You do not know, unless you can get at a vein, that the vein is swollen. You cannot know unless you can get at a liver that the liver is swollen. You cannot know at all if you have a vein or an artery that's in this aneurystic state up in the brain. So high blood pressure is intrinsically dangerous because you are pounding stuff into a pipe in which there may be, and there probably are, a lot of little microburst aneurysms that occur quite often. When you're bruised, your bruise is a capillary breakage in the body and that discoloration is little micro capillaries that have broken and bled into the tissue. They are reabsorbed and the body repairs. But large ones can't reabsorb because there's too much blood loss.

So you bring down blood pressure. Here's a drug that will drop blood pressure. This is one point I wanted to bring up. I was asked by someone in the last lecture for an example of a sympatholytic drug. The example I gave was Haldol. It's not a good example. It's more of a dopaminergic drug. But a good drug would be, since the sympathomimetic urge is to increase the fight and flight capability, the sympatholytic one that would decrease it, such as dropping blood pressure, a lot of the drugs that are given to drop blood pressure are sympatholytic drugs.

So you want to bring down blood pressure just to protect the pounding of the pipes. And you do it by giving a drug that interferes with that sympathomimetic action. Remember the idea of a stimulant: dilated eyes, running away from the saber-toothed tiger, saving your life, moving, not digesting, maybe peeing in your pants, but the main thing is not stopping and going to sleep and eating and digesting and making love, but running and fighting and getting out of there, where of course blood pressure is a part of it. The heart just really begins pounding. You want more blood, you want perfusion to the muscles, you want oxygen out where oxygen's needed.

So the anti-sympathomimetic, the sympatholytic, is a way of dropping blood pressure. And it's used if you have such and such a blood pressure, it's too high, let's bring it down. This is an example of the treatment of an ill person to keep them ill, to suppress the symptoms. You are hiding the blood pressure, you are not repairing the blood pressure. You're not doing a single thing to improve that illness. Just like giving insulin does not build a new pancreas, and giving a thyroid extract does not make a new thyroid. So really, you are treating the symptoms and suppressing them. Cosmetics. You're covering it over.

Eliminating the symptoms requires an aggressive approach because most symptoms are caused by something being there that's aggressive: a bacterium growing where it shouldn't grow, a virus reproducing like fury when it shouldn't be there, a fungus that grows between the toes when it shouldn't be there. Something, as a rule, is the aggressor to produce illness and your way of treating that and getting the person well is to attack the aggressor. So you need really a disruptive material. A bactericide ("cide" or "cidal": to kill) is something that kills bacteria. A bacteriostat is something that keeps bacteria from growing; keeps it static. So you can have a fungistat or a fungicide, or a bacteriostat or a bactericide, or a virostat or a viricide. But the idea is to get at the organism that causes the disruption, get rid of the organism, the person gets well because the organism is no longer spewing its toxin into the body or reproducing or blocking something or causing something.

I just suddenly thought of another example of suppressing the symptoms, a fever. You run a fever. What is the function of a fever in the body? Well, it may be to drive the temperature of the body up so that some organism that can't take the higher temperature can't reproduce as readily and is gotten rid of. You, being much larger, have more reserves and are better equipped to survive the heat. But fevers can be damaging. And if you are running a fever for unknown causes, one of the first moves a physician will make is give you something that brings the fever down. Just like if you have a pain the doctor will give something to get rid of the pain. Bringing the fever down does not address the cause of the fever, just like getting rid of the pain does not address the cause of the pain. These are symptom suppressors. Too much fever, you will succumb to it.

I have an example, in San Francisco at the medical center about eight or ten years ago in dentistry—I happened to be in the vicinity and was in on it a little bit—a woman about twenty-one years old went into surgery, elective surgery, to remove two impacted molars, but they were not in distress. The old saying, "Surgery can't be casual, there is no casual surgery." Believe me, anytime you are in any surgical intervention there is a sometimes small risk, but there is a risk. She went into elective surgery to have the molars removed. Went through absolutely the proper protocol. There was no deviation, no error made. There was a muscular relaxant given, which is very common presurgical technique in which you somewhat paralyze the body. The idea is to eliminate the reflexes to some extent because you want—I don't know if I'll get into this later, so I'll get into it right now.

You want in surgery, where you give a general anesthetic, to use as small an amount of drug as possible. You always, in all medicine, all treatment, all intervention, where you are disrupting the body, use as little as you can get away with. In chemistry if you're extracting something with a solvent, use the worst solvent you can use if it's good enough. Don't use the best solvent, you get too much. In medicine, use as little a drug as you need, use as mild, as non-toxic, as inoffensive a drug as you can get away with. With an anesthetic, use as small an amount of anesthetic as you can. So since the body's reflexes often get in the way, you cut out the twitching (you may not feel it, but you may twitch) by giving a paralytic. A person's all uptight because they don't want to go into surgery, don't want to go under the knife, you know; give a thing to make them less uptight. Give them a drug to give them a little bit of amnesia so they're kind of wishy-washy about going into surgery. They're not going to fight you. At which point you can give a very narrow, very low level of anesthetic and get a person into a surgical area where you can do surgery with a certain amount of depth.

And that's exactly what was done. Four drugs were used: something to somewhat paralyze the body, one for tension, another one for anxiety, and something to generally dry up the mouth. The last thing you want to do when pulling teeth is have saliva flowing everywhere; so you give a parasympatholytic (we'll reemphasize these at the end of the hour) that will dry up the saliva and quiet the gut. You don't want vomiting or the bowels to move. You want the whole thing to be kind of quiet in the middle of this so you can get in there, pull the teeth, and let the person come out of it.

Somehow, at the beginning of the anesthesia her temperature went up from just below normal to begin with, about ninety-eight point six, I think it is, but in centigrade it's thirty-seven. One minute in surgery it went up to thirty-eight. Two minutes in surgery it went up to thirty-nine, she's about now 101 [Fahrenheit]. Fourth minute in surgery it went up to 102 or 102.5. The fifth minute it was 103 and still rising and everyone panicked. The temperature was out of control. The energy of the body had been decoupled, the phosphorylation probably had been decoupled. The body had lost control of its thermostat. And at a degree a minute. She died on the table at about 110. If you knew it was going to be happening, there are drugs that are dramatic that will bring down the temperature just like that, but they have to be at hand. You don't go down to check them out of the pharmacy while someone's temperature is rising a degree a minute. And she died, she died on the table. Something went completely amiss during that surgery and it was never talked much about because malpractice has become a frightening thing. It's a one chance in, god, I don't know the given number because it's kind of frightening, but maybe one in a few thousand, maybe one in a few tens of thousands, in which some drug or some combination of drugs is wrong for that person.

God, I never got off the first page of the handout. Okay. There's a term polypharmacy, which means "many drugs." You have, for example, before a drug can be used in a hospital or as a prescription drug, this mountain of animal toxicity. They've tried it in eighty-five species, through half a dozen generations. They've put it in the weirdest places you can imagine. They've tried it in lots of people to find out what the feedback is. And finally you get this feeling of safety, which you cannot prove, but you can get confident of. And then this drug now can be used. So suddenly this component is used in the hospital. Then another drug that may be for dropping blood pressure, increasing urine flow or removing spots in the eyes, I don't know, whatever, is then added to it, and the mixture of drugs often never has been looked at in any study as a mixture. So the combination of two drugs is really something quite different than either drug alone. And the multiple addition of drugs to a person is called polypharmacy.

A survey was made in New York at one of the major hospitals, I don't remember which one, about ten years ago in ICU. ICU is the intensive care unit. It's the place that is the location of people who are in extreme, critical, minute-by-minute, attention demanding need. This and that is just on the ragged edge of not being in control, or requiring titration in some very delicate way. You have a one-on-one attention in ICU and it's for intensely serious cases who may or may not survive. A survey was made through ICU at one of the major hospitals, about twenty or twenty-five patients, on, at the time of the survey, the best estimate they could make of the number of drugs that were inside that patient. You have this, there was the residue of the anesthetic, you have that because there was a complaint coming in, you have that because it's a drug that the person in there needs for being poisoned against, you have that for something else. How many drugs are in the average person? The average number of drugs was sixteen. There is an extreme unlikeliness that any two of those drugs had ever been tested in any animal together. The probability that sixteen of them had ever been tested together is vanishingly near zero.

So you have situations in medical care where you have mixtures of drugs that are not known. Some mixtures are very, very straightforward, very obvious, very well known. One drug I'm going to put up here in great big letters because it mixes with very few others. I may have mentioned this before. Alcohol is a depressant. I mean, you get your giddiness and you have a nice sloppy looseness from it, which is neat. But then everything begins getting more and more depressed. And alcohol used with a drug that depends upon neural integrity, or a drug that itself is a depressant, often adds together in a way that is more than the sum of the two. There are two words in here. (Whoops, I didn't put them in the definitions I handed out. I'll add them but I have no way of getting them to you.) Potentiation and synergism. Very valuable words. And yell if I misspell [writing on the board].

Potentiation is where one drug makes another drug more active than you would expect. In potentiation, the chemical itself does not have an action appreciably. But it promotes the action of another drug, it potentiates its action. Synergism is where two drugs have their own action, but together they add up to more than total. Two plus two equals five is a good analogy. In this latter matter, one of the most frequently encountered drugs, and without doubt the most dangerous one, is alcohol. It is a depressant in the classic sense of the term although the initial inebriation provides a disinhibition that has been universally used to justify is use. There will be one entire lecture given exclusively to the subject.

As a depressant, alcohol is synergistic with most other depressants, even if there is no immediate chemical resemblance at all. An illustration can be taken from the practices in a clinical laboratory. Let's say you're running the level of alcohol in a person; or, more to the point, let's say you have a person whose blood has just been sent to you for a toxicology screen with the STAT command. This is someone who's an emergency and their blood has been sent to your clinical laboratory. "Help! What's in the blood? What's going on?" It's called a "STAT" situation (from the Latin, *statim*) and means urgent. "Drop everything and get at it, we have a person in a coma, do what you can to get any clue." Get any information you can, as fast as you can.

This is usually in conjunction with some life-threatening situation that might require different intervention approaches depending on the diagnosis. One of the first things you look for is barbs. You find a big level of barbs, you get on the phone and say, "Barb overdose." And they'll say, "Find out what barb it is; we'll start treating for that." Maybe they'll intervene with an antidote. And maybe you have to dialyze a person and maybe you have to support, most emergency treatment is support anyway. If something goes to the right and it shouldn't be there, you put it back to the left. You don't go in and give specific antidotes unless you have a clue what's going on. You can do more damage than good. So most emergency treatment is life supporting. And then it becomes specific when you find out what drug you're handling or what situation is failing.

But, if you find a modest amount of barb, automatically, anybody in a clinical lab will run a blood-alcohol test. Because a small amount of barb is pretty much a normal level and by itself not critical. And a small amount of alcohol that is not normal, which I guess is normal in our society, represents a couple to three drinks, with a small amount of barb, the barb you would normally take to go to bed, together this can be lethal. It's more than two plus two. It's two plus two equals about eight or so. It's a case of synergism. Alcohol is synergistic with a lot of things that are depressants; not just barbiturates, but other sedatives such as Doriden or methaqualone, or even benzodiazepines such as Librium or Valium can be potentially lethal with alcohol. And we happen to use a lot of depressants in our society for philosophic reasons that we may get into later. But anyway, the combination is not a good combination. That is called synergism.

Okay, back to where I should have been twenty minutes ago.

ANN: Is it possible to have a synergistic reaction between two drugs that actually is something that you want? I mean, can you deliberately use one plus one to equal three or four?

SASHA: Yes. This is done sometimes where a person becomes intolerant of a drug because of idiosyncratic response, so you want a smaller amount of the drug that's an ineffective level. You can sometimes use two drugs, both at ineffective levels, both being able to be tolerated as individuals, but they sum to an action that you want. This is done in the anti-epileptics; often we use mixed drugs. In the area of anticonvulsants, a commonly used barbiturate is phenobarbital. Besides the above mentioned problem with alcohol, the chronic use of phenobarbital will lead to a process known as enzyme induction. In the simplest terms, the continuing use of a drug such as phenobarbital changes the metabolic activity of the liver to process the drug. So with time, the efficiency of the drug will change due to this continuous exposure. This change in metabolic capacity will also affect any other drug that calls upon the same metabolic system. Dilantin, another anticonvulsant, is similarly metabolized, and in a person who has been chronically exposed to phe-nobarbital (as opposed to a person free of it) a given dose of Dilantin will be radically different in its effectiveness.

Mixed drugs are looked on with a jaundiced eye by the FDA because mixed drugs are very rarely approved as that mixture. But a lot of things are drug mixtures on the market. You go to over the counter sales in a drugstore, pick up a box of something that treats this or treats that or treats something else, and look on there, it contains this, this, this, and this. And that's a drug mixture.

Barbs can give an example of, not a synergism, but a drug mixture. The barb Tuinal is a mixture of amobarbital and secobarbital, one of which is rather fast acting and drops off fairly quickly; the other is slow onset, but lasts for quite a while. And the mixture is used so you get both early sedation and extended sleep. So you're in essence taking two drugs, one fast and one slow. Another example of a mixture would be drugs with opposite action (such as Dexamyl, a sedative and a stimulant) for mutual modification of side-effects. In the medical community, in the AMA [American Medical Association] listing of drugs, they're called "irrational mixtures" because often they're mixed one to two or one to one or some unit combination where that may be the wrong thing for a given individual, but it's the combination that seems to strike the best average. So it's irrational because they're integers, they're not tailored for the individual. You wish to usually use a drug to the extent that the individual needs it. And when you use a mixture, you may be right with one, but wrong with the other, so the mixture is called "irrational."

STUDENT: Tuinal has a high addiction potential.

SASHA: I would put that generally all barbs at certain levels have a high addiction potential. Tuinal is certainly in there for addiction potential.

ANN: This would be true physical addiction?

SASHA: True physical addiction, yes.

Okay, back to our four-cornered grid, the upper right hand corner is drugs that are used by healthy people to maintain health. The most conspicuous examples of that would be things like inoculations, vaccinations against certain things, treatment with preventative prophylactics. We mentioned the term prophylactic, building a shield, or building some sort of a defense against something. I will hold that in an anatomical sense, pregnancy is a pathological state that borders on cancer. In this case you are treating a healthy person to keep her healthy. If you take a look at the concept, suddenly there is an alien chromosome present; after all, the sperm has nothing to do with *your* genetics at all unless you're really deeply inbred. Once in the body it couples with a cell and the cell becomes a thing that becomes two cells, becomes four, becomes eight, becomes sixteen, and a whole bunch more. And it divides and divides and divides and doesn't carry your genes. It carries part of your genetic system, but it doesn't carry the rest of it. What's the definition of a cancer? Something that's multiplying out of control again and again and again that's alien to the body. A developing embryo (the term embryo and fetus, by the way, should be held separately) is usually just for the first couple to three months where the cell is dividing until it becomes the shape of the organism that it is. That is the embryonic stage. The organism may be only that large [gesturing to indicate something small], but it's got a head, it's got arms, it's got the buds for all the limbs. Then the growing organism is known as a fetus. So you have the embryonic stage of development and the fetal stage of development. The break point is around two or three months.

But this is an organism that's an alien thing! I mean, good God, the body is going to want to reject that. And boy, if it doesn't reject it in about nine months you've got serious problems of other sorts. [Laughter.] So stop and think, why doesn't the body reject all developing embryos as soon as it picks them up? You've got a beautiful immune system. A woman should have an immune system that says, "There is an alien thing growing. Reject!" Well, most are. A lot of implanted embryos are rejected. Miscarriages occur far more often than they are apparent.

But the thing is, somehow that uterus has a way of screening the system outside that says, "Here are my antibodies. We're gonna get rid of foreign things," from the developing embryo inside. It sort of hides the information, it's kind of a Berlin Wall, that on the other side of that wall there is something that's alien.

This is being looked at very keenly because what a beautiful way it suggests for getting at cancer. If you could somehow not hide from the body the knowledge that the cancerous growth is alien, you could locate it, you could direct things to it. There are a number of tools toward cancer called cancer recognition, taking advantage of the fact that a lot of cancers develop very quickly. They develop so quickly that the vascularization to the cancer cannot develop to keep up with the cancer. And the cancer grows and it gets dead inside because it can't get blood to it. It's growing too fast. It's dead inside, it's cool inside and there's not much metabolism going on, because the air can't get there. You have now, for example, drugs that are being developed that will seek out, that seem to attach themselves to, cell tissue that is not using enough air, that is not aerobic. It'll go in the body and get inside of a cancer and attach itself, but the drug carries with it something that will intercept radioactivity. It will be opaque to radioactivity. So you can, in principle, give this drug, it'll go to the inside of a cancer because that's where the cells don't use that much air, they don't vasculate quickly enough, and then irradiate the body with the main radiation being effective for that cancer. These are the kinds of approaches are being used. Taking advantage of this abnormal cell development.

But in the development in the uterus, you have an abnormal thing and it is like, in essence, a cancer. In that sense, the use of a thing that prevents conception could be considered as a thing given to a normal, well person to keep a normal person well. It's a rough kind of classification, especially in certain enthusiastic points of view toward abortion and anti-abortion, which I am totally going to stay away from. [Laughter.] Anyway, the idea of contraception, which a lot of people want to stay away from the concept of, and unfortunately a lot of them stay away from the employment of [contraception] because of ignorance or haste or God knows what. When I read about the number of teenage pregnancies, the number is absolutely astounding, and it far exceeds the extent of lack of knowledge about how to prevent it. So the idea of more education in that process may or may not be a virtuous thing. I think the same way of drugs. In many ways the issue of drug education is parallel to that of sex education. Much of what has been learned has come from one's peers. The parents expect that somehow it will be taught at school, and the school is afraid that if there is an attempt to really cover the subject adequately, there will be some objection from the parents. "Just say no" has its role here as well. Abstinence may be good advice, but what about the person who chooses not to be abstinent?

I think the amount of drug use is rather astounding in spite of the fact there is some education that has been given at some levels about what is good and what is bad about certain drugs. So I think there's a certain amount of drug use that's going to go on despite education, and a certain amount of sexual interplay that's going to go on despite education. So I think we'll have pregnancy and drug overuse as a mainstay for a long time.

[Directed to student] Yes.

STUDENT: Do you mean the education that teenagers and young children receive about drug use is inaccurate up until college level?

SASHA: Why do you draw the line at college level? Because even there sometimes—no, a lot of it is woefully inaccurate. Even the quantity of information that is taught in the medical schools to doctors-to-be is limited. A lot of it is accurate, but it's this much of the story [gesturing]. And the rest of the story is also accurate and it's not given. I mean, I got a question just a couple days ago after the last class. Someone asked, "What is the source or what is the truth in ecstasy depleting the spinal fluid?" The drug MDMA depleting the spinal fluid? I'm aghast! Never heard of any such thing. It's absolute nonsense. And yet it is one of the stories that's being put around.

[Directed to student] Yes.

STUDENT: So you think that kind of education, incomplete or inaccurate education, may promote drug use as opposed to—?

SASHA: Absolutely! I think it is extremely counterproductive. I mean, look at the gut response of a lot of people when you have something coming out in the newspaper, "Doctor so-and-so at the National Institute of Mental Health says we now have evidence that such-and-such a drug will increase the tubular diameter of the cerebral veins in rats." Say marijuana has been shown to do this and this and this. They've got an experiment where they have shown some negative thing about marijuana; further evidence that people shouldn't use marijuana. How many people who use marijuana are going to suddenly say, "Oh my god! These rats blew up their cerebral ventricles on marijuana. Maybe they're right! Maybe I shouldn't!" [Loud laughter.] Or you get this sort of thing when suddenly a new drug appears on the scene—I wish I had brought it, I found an editorial, of all places, in the JAMA, Journal of the American Medical Association, that talked about "designer drugs." And these are exactly the kind of pejorative, meta-message words that are slipped in. They're talking about fentanyl and methylfentanyl. We're going to get into them when we get into narcotics. Sooner or later we're going to get to drugs. We're still laying some ground here. [Laughter.] But, fentanyl is a very, very potent morphine substitute that is used in surgery. It is a central, meaning it operates in the brain, analgesic, in that it means you do not respond to pain. You may feel it, but you don't respond to it. So if you're unconscious and have fentanyl, it's a nice adjunct to anesthesia. You have alpha-methylfentanyl, which is a compound with a methyl group tucked on here. Fentanyl, illegal, alpha-methylfentanyl, not illegal as of that time. Now, emergency scheduling has changed that, but now there's beta-methylfentanyl that's not illegal. The idea of designer drugs is you can put a tick or a tack on a molecule, not change its pharmacology greatly, but change its legal status totally. This is the concept of designer drugs. We'll have a whole lecture on it later.

But the editorial said that fentanyl is a safe narcotic used in surgery (70 percent of all anesthesia now uses fentanyl) that is a jillion times more potent than morphine and it, along with meperidine, are the most abused drugs by medical professionals because they're readily available and they are narcotics and a lot of people are very much into the narcotic scene, including physicians and including nurses, as well as people who walk along on Market Street holding you up. [Laughter.] There are a lot of people who are into narcotics. Okay. Fentanyl is a safe narcotic that is such and such. Methylfentanyl is a dangerous narcotic because it is this and this and this. But their reasons were the same. They cause narcosis, they cause the very same results.

This one is made by a pharmaceutical house for use in a hospital. This one is made in an attic in a run-down spot in the Haight-Ashbury for sale down in Golden Gate Park. But the molecules are the same as far as their hazard to the body. They both cause the same things: the pinpointing pupils, the narcotic response. Why is this safe and this dangerous? That's absolutely ridiculous. They're both equally dangerous. They're both equally safe. Depends on how they're used and what they're used for. Then they suddenly come out with a new drug and they say, "This drug in preliminary studies is shown to be this, and is worse than that, and is as bad as this and worse than that, and we absolutely will drive it into an illegal category because the abuse potential is very high, and it has no medical utility and all use is bad." And people will say, "Oh, whoa, they said the same thing about the last one, and it turns out they were wrong." It's this idea of crying wolf so often that it's not listened to. And I can make a little collection that I call "cry wolf" in my file cabinet at home, new drugs that are talked about in the paper and suddenly you see this absolute projection of everything that is evil about what you don't like about certain drug use suddenly dumped upon a new drug.

You have this now. THC has worked its way, in a very strange way, into a Schedule II classification from Schedule I because it is used in certain cases—in fact, this is a major trauma in the whole legal system—for the treatment of nausea associated with the radiation therapy of people with cancer. They've known for years that some people can go through intense nausea situations, with marijuana to suppress that nausea. It is a superb anti-emetic, anti-nausea—what would be the correct term, I guess anti-nausea would be the correct word for it. And so, they finally said that in experiments they found that THC in a capsule taken with radiotherapy will soften that amount of nausea.

So they removed THC from Schedule I. This, by the way, took a number of years to get through the paperwork, and they were fought all the way by, of all things, NORML, the National Organization for the Reform of Marijuana Laws, who didn't want THC taken out of Schedule I and put into Schedule II. NORML wanted ALL restrictions on marijuana to go into Schedule II. And those wanting only THC to be approved didn't want all to go because that would give the message that smoking dope is good, or that we approve of smoking dope. Just like we don't want to give free hypodermic needles to people who keep shooting up with heroin because it gives the message that shooting up with heroin is okay. Far better to have hepatitis and AIDS than to give a message that it's okay to shoot up with heroin. So you don't have free needles available, you don't want pot to go from Schedule I to Schedule II.

So they took THC out [of Schedule I] and they said, "Only if it is put in a gelatin capsule along with sesame oil and it's only prescribed for use by people in radiotherapy for cancer in cases of nausea. It is not amenable to other treatments." Nausea for chemotherapy, that's it. And suddenly the physicians are up in arms, "What do they mean telling me how I can practice medicine? I can only use the drug for this specific use? Some legislator, some guy writing regulations in the DEA is going to tell me how to practice medicine? Nonsense!" So the FDA is out of shape, the medical community is out of shape, NORML's out of shape, everyone's nose is bent because they have taken THC from Schedule I into Schedule II, and it's taken six years to do that. And now other people say there are other drugs that are just as good and it shouldn't have been done. But the whole situation is because THC and marijuana have been invested with every single evil. They are used as a dumping ground.

We have alcohol and tobacco and lots of other drugs that are very clear public hazards. How much public hazard is associated with marijuana that may be quite real? But meanwhile, it and PCP and heroin are the Schedule I targets for all the projections of evil. It's like the Russians are targets for our American community and we're targets for the Russian community.

[Directed to student] Yes.

STUDENT: THC was considered classified as an analeptic?

SASHA: It's classified legally as a hallucinogen.

STUDENT: But, I mean the medical use for the nausea, isn't that for an analeptic?

SASHA: Dysleptic is interruption of the nervous system. Analeptic is around the edge of the nervous system. It's one that, in essence, stimulates the nervous system.

ANN: What happened with the use of THC?

SASHA: Oh, it's still very valid in research, but it's not been legally changed.

ANN: They can't use the THC except in sesame oil.

STUDENT: So it has to be in a capsule.

SASHA: And used for this one purpose, which is the dictation of medicine by law, which the law says "we will not do" and they have done. So that's still in the throes of turmoil right now.

[Directed to student] Yes.

STUDENT: But, taking the capsule is not as effective as smoking it?

SASHA: No, apparently it's not as effective.

STUDENT: It takes longer.

SASHA: It takes longer and it's in oil so it is harder to be able to regulate dosage. Smoking has the advantage of immediate delivery to the bloodstream and you can titrate the effects immediately. Smoking has the disadvantage that a surprisingly large number of people do not know how to inhale. And you cannot go to a non-inhaling, non-smoking person and have them smoke. The mechanisms are just not known. Probably the best way would be intravenous administration, which you'd get directly in, but that is not the form in which it's been approved. Sesame oil intravenously is not considered nice. [Laughter.]

That is the third category where you give anesthetics. If you take a person who is in a reasonably good mental state, and reasonably attentive, and has faculties in good shape, and lay them out on a table and put a mask over their nose and give them ether for a few minutes, they're going to go into a very strange altered place. They're going to go into a state from which they will have their own dreams and cannot communicate. You can open them up, you can remove things, you can put things in. They do not respond. Here's a case where appendicitis may not be a very good thing, but the person's mental and psychic state is good to a large measure, which you are destroying the goodness of by an anesthetic. So an anesthetic is something that's given to a well person to make them unwell: non-communicative, non-feeling, non-responsive, which is not the normal state. So anesthetics represent kind of a kosher type of lower right-hand corner things given to well people that are disorganizing.

Things given to well people that are disorganizing, not so much in the kosher area, but more in the altered state area, are a lot of the things that we will be talking about at some length during the entire course of this semester. Drugs that cause change in perception, change in attitude, change in point of view, sometimes change in conscious integrity, sometimes change in intellectual integrity, visual and sensory changes. These are the psychedelic, or hallucinogenic, drugs, which will be covered in some depth. A broad term that has been used to cover drugs that can modify and influence brain function is "psychotropic," meaning the turning of the mind. We are going to go through a great deal of this.

I put on the board a collection of five categories from the work of Louis Lewin, a pharmacologist/pharmacognosist around the turn of the twentieth century, and the first person to bring peyote into Western research. He brought it back from Mexico in the 1880s after being made aware of it by Parke-Davis. He went back to Germany and isolated a number of the alkaloids from it, although not including mescaline. That was accomplished a few years later by Arthur Heffter. Lewin was one of the earliest reporters of the psychopharmacological scene and recorded much of the social use of drugs, for example, ether and chloroform.

I think I mentioned some of this going through the history of drug development. There is a term "etheromania" for people who really got carried away. When ether was first isolated it was called sweet vitriol because it was the result of sulfuric acid on alcohol steam distilled as a separate layer that was quite sweet. Taste ether, it has a distinct sweet taste to it, and a very sweet smell, and is a very good intoxicant. You take a half teaspoonful of ether and swallow it, it's going to burn like hell, but boy, you're drunk! And when ether first appeared, they had ether parties, this was back in about 1820 or 1830. Around the university, somebody would make up ether, they'd put it in parties, they'd inhale it, they'd take it from big canisters, they would swallow it. It was used as an intoxicant.

Same with nitrous oxide when it first came on the scene. People would make touring caravans around the East Coast, "Nitrous oxide! Come to a demonstration this evening. Thirty gallons will be made available to the people who participate. See how silly you can be, how silly others can be." It was a novelty, a caravan type of a carpet bag job. And many drugs when they first came out had this whole aspect of usage, something that's not totally different from today. And it resulted in etheromania in this country, largely on the East Coast, and in eastern Germany and throughout the central part of Europe. Another chemical used in the same area, that was used in the same way, was chloroform, when it first was evolved, because it has a sweet smell and it caused a disruption of consciousness. These were the drugs that Lewin was studying and recording, along with peyote and mescaline.

And those who always laugh, now laugh the more.

A GRAND
EXHIBITION

OF THE EFFECTS PRODUCED BY INHALING

NITROUS OXIDE, EXHILERATING, OR

LAUGHING GAS!

WILL BE GIVEN AT *The Masonic Hall*

Saturday **EVENING,** *15th* **1845.**

30 **GALLONS OF GAS** will be prepared and administered to all in the audience who desire to inhale it.

MEN will be invited from the audience, to protect those under the influence of the Gas from injuring themselves or others. This course is adopted that no apprehension of danger may be entertained. Probably no one will attempt to fight.

THE EFFECT OF THE GAS is to make those who inhale it, either

LAUGH, SING, DANCE, SPEAK OR FIGHT, &c. &c.

according to the leading trait of their character. They seem to retain consciousness enough not to say or do that which they would have occasion to regret.

N. B. The Gas will be administered only to gentlemen of the first respectability. The object is to make the entertainment in every respect, a genteel affair.

Those who inhale the Gas once, are always anxious to inhale it the second time. There is not an exception to this rule.

No language can describe the delightful sensation produced. Robert Southey, (poet) once said that "the atmosphere of the highest of all possible heavens must be composed of this Gas."

For a full account of the effect produced upon some of the most distinguished men of Europe, see Hooper's Medical Dictionary, under the head of Nitrogen.

The History and properties of the Gas will be explained at the commencement of the entertainment.

The entertainment will be accompanied by experiments in

ELECTRICITY.

ENTERTAINMENT TO COMMENCE AT 7 O'CLOCK.

TICKETS *12½* CENTS,

Lewin was the one who coined five general categories for things that affect the mind. It's kind of quaint. He put them in Latin just to make it a little more international. *Excitantia,* things that excite. Remember I mentioned the ups, downs, and stars of the DEA's handbook on stimulants, depressants and hallucinogens? Excitantia is perhaps the "ups," the stimulants. You find them all through nature. In Lewin's time, there were very few synthetic chemicals that were known to have drug action or even thought of as drugs, and most of the materials looked at were from natural sources. Of Excitantia, we have many stimulants and excitants in nature. We mentioned caffeine, one of the major ones. But you also have ephedrine. You have khat, a material I may talk about in much more detail, a plant that grows in the Asia Minor area. It's very much like ephedra, but it has an amphetamine skeleton with a carbonyl group. It's quite a different molecule. It's used, it does not store well, but it is a stimulant. What other stimulants do we have? The lesser purine alkaloids have stimulating action.

Inebriantia. Things causing intoxication. Things that cause confusion. Alcohol is one of the best examples. Many of the drugs that were explored then would cause sleep and smaller amounts would cause this excitement, this inebriation, this intoxication first. Going back to alcohol, probably one of the classic examples; at a certain level of alcohol you begin losing muscular coordination, losing neurological integrity. You will eventually end up sleeping or in a coma. The cause of death by alcohol is almost always asphyxia. It's not due to the fact that you have enough alcohol in you to stop things, it's that you get into a coma and you vomit. Vomiting when you're unconscious is one of the most hazardous acts because the vomiting is a convulsive reflex and it's not uncommon when you vomit to gasp. Gasping is part of this process. If the vomitus is out or coming out, if you gasp in in any way during that process you haul the vomitus into the lung, and that is where your damage comes. You can actually block air access.

There are four general categories in which you never cause a person to vomit in overdose. This is a good place to put this in. If a person is having a drug problem, the first thought that often goes through the mind is, "Get the goop out of the tummy. Get the stuff out of the system. Vomit." Never, never have a person vomit if the person is comatose or unconscious. A finger down the throat will often trigger it, and often that reflex will be effective, but if a person is comatose or unconscious, that gasping could very well do more damage. It goes right back into the lungs and you have a very serious problem. Better leave it down there.

Never cause the person to vomit if the person has taken a hydrocarbon. Let's say a person has been siphoning gas out of someone's tank to put it into someone else's tank, or whatever one would do to get hydrocarbons in the mouth. Siphoning gasoline is a good one or maybe swallowing cleaner fluid, a child who swallowed cleaner fluid because it's been stored in a Coca Cola bottle. I can give a whole editorial along this line on how you store poisons in the home when there are small children around; Don't put them in bottles that say Coca Cola or are known to contain nice things. Consider a person swallowing cleaner fluid, a hydrocarbon; the problem with vomiting with it is they are going to get hydrocarbons in the lungs. Liquid, particulate, aerosol hydrocarbons. That's bad. With gaseous hydrocarbons, some may be absorbed, they'll be exhaled, but liquid hydrocarbons get into a certain particulate size and they can't get out of the lungs. Big things are carried out by the villi that continually move carrying scrunge out of the lungs. Very small things get absorbed into the body. But there's an in-between size that a lung can't do a thing with. Sometimes the aerosols or certain sols or certain particulate dust will get into the lungs and can't be handled. You get very difficult problems from that.

Never have a person vomit if the person's taken a caustic, a corrosive. Let's say a person's taken a big pile of sodium hydroxide. Take it out, if you can, with a stomach pump. Neutralize it in place with an acid, but don't have the person vomit.

[Directed to Ann] Yes.

ANN: Is that lye?

SASHA: Lye or sodium hydroxide. It may be by accident, it may be by intent. But what it does is to corrode, erode, caustically destroy the tissue to a large measure going down. I've seen one case of a person who survived a suicide attempt with lye, and that person had a continual problem with opening up the passageway from the mouth to the stomach, continually, because it was tending always to fuse together. That lye will lie like a lump in the stomach. If you vomit it out, it does double damage because it hits the tissue again coming out. Better to leave it there and treat it in place. Vomiting is not the best treatment for lye swallowing.

[Directed to student] Yes.

STUDENT: Is that also true because the acid in the stomach is acidic and you have a basic substance?

SASHA: The acid is long since gone. It's been neutralized by the lye. So they stimulate and put more acid in. The attempt is to neutralize it there. And the best thing is to remove it mechanically by lavage, which is stomach pumping, or by adding acid.

[Directed to student] Yes.

STUDENT: What do you do to neutralize it?

SASHA: Add acid. Pump it out or add acid. This is the medical intervention. Either one. You do not have a stomach pump in most homes. [Laughter.] So emergency intervention.

The fourth time you never cause a person to vomit is if the person has taken a convulsant. Now, you don't always know what this is. In fact, half the problem of emergency medicine when treating overdose is you don't know what went down. Some compounds you can smell. Some of them you can tell by conspicuous signs and some you just don't know. But if you find an open bottle of strychnine alongside the person, then you have a pretty good idea the person is not moving much because of a strychnine overdose. The very act of vomiting is an act of convulsion. Causing a person to vomit will trigger the whole convulsive syndrome. So, don't trigger vomiting with strychnine. There are other ways to respond, such as lavage, or you can get carbon in them.

Those are the four overdose situations where you don't want to induce vomiting. If you do want to cause vomiting, there's one thing everyone should have and that's ipecac. It's available without prescription in a drugstore. The directions are on the bottle. It can be useful if you have small children around and you realize one of them has gotten into something and you want them to vomit. A spoonful of ipecac will almost always produce vomiting within a very few minutes. If it doesn't, a second spoonful almost always will succeed when the first one failed. If it doesn't work after two tries, don't try again. Something's wrong.

[Directed to student] Yes.

STUDENT: Doesn't it have potential abuse? Karen Carpenter supposedly was using it for her bulimia and anorexia, and it was one of the reasons why she died.

SASHA: Yes, it does. Back to the very first lecture. There is no drug without terrific potential for abuse and misuse, and there are very few drugs that do not have good and jolly virtues attached to them at some level as well. It's a matter of how you use it, why you use it, and how much you use, and what you and your relationship to the drug is. Let's go back to ipecac. How many people are aware of the term "bulimia?" They have a literary way of expressing it.

STUDENT: Binge and purge.

SASHA: Binge and purge. The compulsion to eat and then to vomit and then to eat and then to vomit. What is the illness where people have the image that they're too fat?

STUDENT: Anorexia.

SASHA: *Anorexia nervosa.* The two are, in a sense, allied in some ways. Both are mostly female complaints and often occur during the teenage and after teenage years. And are really extraordinarily difficult problems to address.

[Directed to student] Yeah.

STUDENT: Is ipecac a spasmodic? What causes the convulsion?

SASHA: Don't know how it works.

STUDENT: Is it a natural product?

SASHA: It's a natural product, comes from a plant. The name is twice as long as ipecac.

ANN: We have a question. What is the danger of having a convulsion?

SASHA: You trigger a grand mal situation, you break your back, you lock up things, you cannot breathe, you can die of asphyxiation, you can die of neurological section. True full grand mal convulsion involves the entire body and can actually break the spine. That's the strength of your muscles, they're very, very strong.

ANN: But, it doesn't damage the brain cells in and of itself?

SASHA: No, unless you cut off the circulation to some of the mechanisms.

Okay, where are we? That was *inebriantia. Hypnotica.* Hypnotica would be the depressant aspect of the central nervous system action. I think I mentioned before that a common pair of words—I'll write them down again. (Wooh! I'm running out of time. I haven't got to what I want to get to.) Sedative-hypnotic. Sedative is quieting, stilling, taking away the nervous reflexia, the tension, the tremor, the anxiety, the "I want to get over this day. I've been hassled all day; I want to be sedated. I want to be quieted down." Sedative is quieting down. Hypnotic is going to sleep. If the quieting down of the drug also promotes sleep then the phrase is sedative-hypnotic. Lewin's term is hypnotica. Some barbiturates are examples. Some of the tranquilizers fit in this category in the synthetic drugs. Very often those things that inebriate at one level become hypnotics at a higher level. Alcohol inebriates and then causes sleep. Chloroform and ether, the very compounds we were talking about that were abused so widely at the turn of the century or before the turn of the century, were also hypnotics. Here also are drugs that cause a dreamlike state, clouded by amnesia, like scopolamine. These are those that are probably closest to true hallucinogens. These lead to a delusional state, confusion, misinterpretation, occasionally vivid spontaneous hallucinations, and all usually masked from recall by a short-term memory loss.

Euphorica is the large classification that now has been gathered under the area of the narcotics. It includes those that quiet not just the outside pain of the physical hurt, but the inside pain of emotional hurt. The relief from stress by the numbing of the awareness of the stress. They are the ones that get you divorced from the trauma by sort of clouding off the access to the trauma. How many people have gone into surgery or gone into some trauma and have been pre-treated with Demerol or with morphine? Most, good. It's not a nice place really, but what it does is to separate you from the intellectual recognition of where you're going. It just doesn't kind of matter that much anymore. For a while, there is no need to cope. The goodies of heroin and the value, the reason heroin is used, and these narcotics are so regularly used, is they cause that anesthesia of the psychic pain. An appropriate image in the ups, downs, and stars vocabulary should probably be an inside-out star—maybe a bunch of arrows all pointing inwards to a point. They cause that inward turning and a little escape into dreamworld and (as I mentioned in the handout, the definition of euphorica, what should be the right feeling, but really the good feeling) the separation from a lot of the trauma of the world that you just can't or don't wish to cope with for a little while. That is the isolation that comes from the euphorica, from heroin, from the morphine type of drug.

And then the *phantastica*, I guess, might as well be the counterpart. It's quite the opposite of euphorica. It's the one that makes you *more* aware, more attentive, to what is around. They emphasize the senses rather than interfere with them. The enhancement can be carried to the point of distortion, but what is seen is more an interpretive distortion of something real rather than the creation of something unreal. True hallucinations are rare, compared with the drugs mentioned within the hypnotica group. These are in the area that often has been called the psychedelics or the psychotomimetics, or the hallucinogens. A synonym of European origin is psychodysleptic, meaning to disrupt the mind. My god, there are dozens of names in this area. This is the group implied by the DEA's stars. I'd like to spend at least one lecture on these when I go back and organize the lectures along the lines of neurotransmitters. These are the materials that enhance the senses, that actually enhance to the point that can distort. Or they can enhance or modify the intellectual integrity, sometimes in a confusing way, but sometimes in a very insightful, and, at least from the person's own personal interpretation and analysis, beneficial way. Objectively, there's a lot of controversy. But from the person's personal seeing of it, very often the use is considered beneficial and that is where, of course, the habituation and eventually the chronic usage comes in. All of these drugs, from the negative to the positive, the ups and the downs and the stars (and the question marks and the pointy inwards in between) are habituating and all can cause dependency. If something feels good, one tends to do it again. If pain feels good you tend to do it again. If something turns you off into a quieted person who drops off to sleep, which is the epitome of escape, if it feels good to you, you tend to do it again. If you like the sloppy disinhibition that comes from not having to be totally responsible for what you're doing and being able to talk a little bit more loosely and not quite as clearly at a cocktail party, you do it again. If you like being wound up and being able to go through the night and drive cross-country, and really get turned on by that kind of thing, you tend to do it again. All these things are habituating and all of them to that extent can cause, and do cause, psychological dependency. Physical dependency, quite another matter. Very few of them do. But most of them will cause a psychological dependency. If it feels good, you tend to do it again.

I mentioned the neurotransmitters in the end of the last lecture. The two that are in the periphery, acetylcholine or the cholinergic system, and norepinephrine or the adrenergic system, these were parts of the peripheral nervous system. I had mentioned the give and take of the autonomic nervous system. A question was asked then about what was an example of a sympatholytic drug. We mentioned sympathomimetic, those that imitate the sympathetic nervous system, clearly your stimulants, your excitantia, fit in that area. Things that cause increased heart rate, things that cause eye dilation, things that cause stimulation, blood flow, cerebral circulation. Things that lyse, that get in the way of the sympathetic branch, the sympatholytic example would be something that drops blood pressure. An example that comes to mind now, in retrospect, would be something like the interfering with the autonomic system that causes the blood pressure itself. Aldomet is a drug that is a good example of the interference with the sympathetic system. An example would be monoamine oxidase inhibitors that will cause the drop in anxiety, anxiety being an expression of the autonomic stimulation of the sympathetic system. A drop in that anxiety would a sympatholytic.

In the parasympatholytics we mentioned atropine, things that actually cause dilation by default, that cause drying up, atropine, scopolamine. There are five signs I think? "Blind as a bat. Dry as bone. Hot as hell. Red as a beet. Mad as a hatter." Your saliva dries up, your gut tends to slow down. It's often a way of pre-treating a person for surgery where you don't want saliva and you don't want gastric motility. The opposite of that would be the parasympathomimetic, things that imitate these things. A lot of the materials that are parasympatholytic are neutralized by materials that in turn encourage the motion of the gut, that in turn act as if they were acetylcholine itself. Many of the insecticides are cholinergic and many of the treatments for insecticide poisoning are anticholinergic. We'll get into much better examples of this when we get into the cholinergic materials such as nerve gases and nerve poisons. I consider things that are chemical warfare agents, such as nerve gases and nuclear bombs and such, all to be drugs in their own way. Since I'm defining drugs, I'll define them as broadly as I want. And I'm going to drag in nerve gases and things that are used in behavior control. They can be chemical, but they can be physical. But in the sense, they are all somehow interfering with the integrity of the body, in that way, they are drugs. So those are probably better examples of the sympatholytic materials.

Most of the drugs that fall into the phantastica classification can be related either to dopamine or serotonin. The two are neurotransmitters that I only briefly mentioned last week. I'll go into each of these in a little bit more detail when I gather the drugs that chemically resemble them into categories. But dopamine and serotonin are both primarily neurotransmitters of the central nervous system. They do not have a function as neurotransmitters outside of the brain, although serotonin is abundant in the gut. If they are given outside of the brain, they don't even go to the brain. They cannot get through the blood-brain barrier I talked about. They have to be made in the brain. Dopamine is a phenethyl-amine, serotonin is an indole, and almost all materials that are centrally active and cause a change in state of consciousness are phenethylamines or are indoles. A few of them would fit in with acetylcholine. They are choline-like, but most of them come into the hypnotica area in that they cause, not a clear change of consciousness that is a positively remembered and sought-after experience, but a change of consciousness that is a delusional thing, that is hidden a little bit in amnesia and much more dreamlike. Just as much a change in consciousness, but not one that you can actively recall, get out, utilize, and exploit.

In the handout, I didn't have time to reinforce it, but I defined some words in the first lecture. Five very, very touchy words are often used mixed up together or used with some uncertainty: hallucination, illusion, delusion, fantasy, and imagery. These are words that are often spun out as being describers of certain altered states. And there is no good definition. Good heavens, no one's going to agree on what a hallucination is. You'll find people will take the smallest thing and the broadest thing and call it that. But I've given here how I use the terms. So, go through the handout in its entirety because I do define a lot of things as how I use the words. You may choose to use them differently. That's perfectly fine, but this is a starting point for trying to unravel some of these highly subjective, emotionally charged words.

The hour is over. We'll get into something new next time.

February 19, 1987

Memory & States of Consciousness

SASHA: Okay, how many do we have? Two, four, six, eight, ten, twelve, fourteen, sixteen, eighteen. Eighteen out of thirty. How does one know the population of a class if the class is never totally populated? [Laughter.]

One thing I'd like to start the day off on today is a subject that was brought up by two people last week, last lecture, and by another person the lecture before, another person just now, and I've got my own share of misgivings about it, and that is the whole area of grades and examinations. I've gotten a feeling that there is a lot of vigorous note taking because there is a lot of collective uncertainty about what is going to have to be done on an exam. And so, better to write everything down, even if you can't understand or spell it, to sort of serve that role as a writing machine. An example, the role of the writing machine. This is an aside. I'm going to continue doing asides for the rest of the year because I happen to think asides, in their own way, tend to sometimes illustrate what you want to say. Sometimes they carry a lecture. In fact, I'd like to make a whole course someday on asides. Have the book, read the book, and the rest of the course would be asides.

This is an aside about when I was in the Navy, during World War II, pretty much in the time before you were in existence, and I was, at the time, an avid bridge player. In fact, I found that bridge constituted much of the art that was missing in poker, and missing in many other card games where you add up, divide, subtract, look in the other person's eyes, and make your bet. In bridge you can actually concoct up a good course of strategy that you could pit against the other person's strategies. You could evaluate their wit or their skill and their background, and you could make a strategy to take advantage of them, but yet not allow yourself to be taken advantage of. I love bridge.

Well the avid bridge player was in the radio shack, a little place where you have radio things, earphones on. He'd always sit there with things going beep, beep, beep. One earphone on and one earphone on the side of his head up here so he could hear the bidding. We were playing bridge and all of a sudden something came over the radio and he was, I think, bidding or whatever it was, and he sort of held the cards over here and his hand was on the typewriter. He must have been the "dummy" or something. He was typing away with this code coming in and we got through the hand and someone else started dealing. And whatever was coming in stopped. He was running about five seconds behind what was coming in. He stopped and pulled the paper out of the typewriter and read it, and for the first time realized what he had written down. He said, "Oh my god, this is an emergency message for the old man. I'd better get out of the game right now and get it up to him." He had served as an instrument, with code coming in the ear, something processing the dots and dashes in some sequence, to pointing to what finger would go where and hit that on a typewriter. And that instrument was a translation machine. He was writing down on the typewriter what came in and had no knowledge of what was there, and went off and did something once he saw what was there.

In some ways, note taking is the same idea. You are somehow running a hand that is running about five seconds behind what's being said and you're desperately trying to take this down, you hope the handwriting is going to be legible, then you realize later, if you ever get around to glancing at it some time before this panicky thing called exams, "By golly, I don't really know what he was talking about there. I don't understand. I hope that that question is not on the exam." In fact, one of the fun things I like to do sometimes in a course is ask people what are the questions they dread most on an exam, and actually use those as the points of my lecturing, which are the points you never, ever get across.

So now we've collected most of the people. We're getting close to two, four, six, eight, ten, twelve, fourteen, sixteen, eighteen, twenty,

twenty-two, twenty-four. That's getting close to almost all. The question, for those who just came in, is: what are we going to do about grades and what are we going to do about exams? And I feel there's a collective uncomfortableness about it. We have a midterm scheduled for a couple weeks down the line, middle of the term. I personally don't like midterms. I think I've already said this. I don't like finals. I don't like exams. I don't like note taking. I like people to have something they can read, they can do it on their own, and listen to what's being said and not to have to take notes. Now, if you have an exam that says "True or false?" or "Which is true?" or "Which of the following five best answers the question?" invariably you have to get into trickiness because if you use a word in there and the person interprets it this way and you interpret it that way, then actually the two answers are right and none are right. You can defend that the two are right, you can defend that none are right, but one is being called for that will go into an IBM machine and the needle is going to go over to sixty-two. So you have sixty-two out of 100 correct. No, I don't like that.

What kind of midterm would people like?

STUDENT: Oral.

SASHA: Oral. Okay, that's a good suggestion. Any other suggestions?

STUDENT: None!

SASHA: None. Excellent suggestion. How about both?
[Directed to student] Yes.

STUDENT: What about a paper?

SASHA: I like the paper for the final. Okay, let's take this oral and this none, which is fine. What I'm trying to do is to get you to stop taking notes, so you're not uptight, and allow me to go on my little manic extensions and examples.

ANN: Can we take fun notes?

SASHA: You can draw pictures if you want. That's fine. But the book is here. How many people have opened the book? Wow! Okay. While we're on the book, I'm going to get, in another two lectures, to stimulants. Before that lecture on stimulants, read the chapter in the book on stimulants. That's Chapter 6, I believe. Read that chapter so when I get in and say, "Any questions? No questions?" we can just sort of go home, which is fine. If there are questions, we can answer them. If there are not, I might think of a few things that I can elaborate upon. But the thing is, the book has fifteen or twenty pages on stimulants. I intend three hours, three hours and a quarter, on stimulants. So I'm going to go a lot more deeply into it, and a lot around the corners, but the book will be a good foundation.

Okay, back to the question of midterms. How many people are uptight about grades? Only one, two? Oh, here come more! How many people don't care about grades? Excellent. How many people are taking this course for no grade? Whoops, I thought that could be done, pass or fail or something like that. So everyone is taking this for a grade? No one here is not taking it for a grade. Okay, grades have some importance. I like the idea you came up with for an oral midterm, or a—what was the other one?

STUDENT: You said a paper.

SASHA: Oh, a paper! I was told when I was asked to give the course, "What you've got to do is this, this, and this. That's what you have to do." This was from the head of the department of chemistry. You've got to have a midterm, a final, you've got to give grades on some kind of a distribution curve, which is nonsense. If I can get you to get the music of what I'm talking about, and some of the philosophies, some of the arguments for freedom of choice in what you do, some of the arguments of not giving away your power to other people who want you to do what they want you to do, this general attitude, drugs being kind of incidental, but a good starting point for this kind of philosophy, and you can see what I'm trying to get at, and you choose your objective, that's lovely.

What I'd like to have you do is get that feel for where I've been and where I'm going, and I think that can be done without taking notes.

So let's try the following. If people object to it vigorously, I'll reconsider it. How about a midterm on the day of the midterm that is oral? I'll go around and ask questions, but here's the cute gimmick: I don't know your name. So there's no way I can grade it. [Laughter.] That I think is a nice way of doing it, in which I'll get feedback questions, "What do you think about this? What is your attitude? Do you agree with me? Disagree with me? Do you remember the relationship between that and yonder?" And just see, to the extent there's English, maybe a couple of the languages, probably other languages as well, but basically in English, and see if you can get a feel for what has happened in the first half of the course. No grades (it's an old thing, I'll probably never be invited back to the restaurant again, that's an old story, but that's okay) and anonymous. That would be a nice way of doing it so you don't even have to prepare for it. Because if you don't know the answer, fine, I blew it! I didn't explain it well enough. Or you were asleep or you were absent that day. That's fine.

Then we get to the ugly part called the final—I do know the convention. When the lectures are over there is a week of finals in which everything is jammed somehow. What a horrible thought. [Laughter.] For one week you are, in essence, in a true altered state. In fact, altered consciousness is what I want to talk about today. So there is a case of a good altered consciousness. You are suddenly being driven by something you can't quite control, you can't know where it is going, you realize the time schedule, you are destined to look at your watch eighteen times a day where normally you wouldn't wear a watch, it's that kind of a week. I don't like that. I don't think it's a good way of evaluating what goes on in a course.

So why not do it totally differently? Ignore the final and I will do the following. I will ask you to write an essay, a legible essay, hopefully. An essay on anything you want to write on as long as it has some rational bearing on what went on in the course. I mean, you want to write an essay on Spinoza, that's kind of neat, but I don't think we're going to get into Spinoza, and it's not really applicable to the course. This is the topic of the essay (I'll reinforce when we get toward the final): Write an essay defending a stance that is contrary to a stance I took. Namely, you disagree with something I came up with, or espoused or tried to explain, where you say, "Hey, hang on a moment. That's not right. I don't agree with that." "It's contrary to my background, my religion, my training, my upbringing, my personal philosophy," whatever. Something you disagree with and you want to take issue with it.

For the final, write an essay on something that you want to take issue with me on, and I'll do the following: If I say, "That is a good position," and you came to me with an argument that is sufficiently well placed, I'll say, "By golly, I hold that as being a correction or a stand I had not appreciated, or something that really bears in on presenting it in that way with more emphasis next time. And I will change what I'm writing to accommodate it." You'll get an A.

If you write an essay on something you disagree with, you think I took the wrong direction, and your defense of your thing is sufficiently wishy-washy or based upon non-factual information (you can take all the time you want and find all the facts you want) and it is not well substantiated, and it doesn't change my way of putting something into the text when I finish writing the text, I'll give you a B. At least you wrote an essay.

If you fail to hand in an essay, I'll give you an F. That sounds fair. There's no point in Cs and Ds because if you've been through the course and you've gotten something out of it, why give you a C? Your grade is something on a grade average of everything from super genius to mediocre genius, and you happen to be average. You may be doing a perfectly fine average job, you get an F, that's ridiculous. This idea of a curve, I don't have any patience with.

What about something like that? And what I'll do, so you don't get into the mess of the final week, we'll do it before finals. Then you can pass it in, let's say at the last lecture, and you'll have no worries about finals in the course at all. You can be all uptight about other courses.

How does this sound? Who takes issue with it? Who agrees with it? More agree than not. [Laughter.]

[Directed to student] Yes.

STUDENT: Does it have to be on taking a stance?

SASHA: Yes.

STUDENT: Can we do it on something else?

SASHA: Take something where I can change my mind about what you present or I will say, "I don't think the argument was presented adequately, or I don't think it's factual."

STUDENT: Yes, but I mean can we write on something else?

SASHA: You can write anything as long as it vaguely associates with the course.

STUDENT: What if we present a strong argument, but you don't agree with it?

SASHA: My agreeing or not agreeing is a matter of opinion. My accepting your arguments is a matter of an analytical approach. I will be honest and do an analytical approach. Oh, there are things that people say are factual, but I just totally disagree. There's a fine difference between truth and fact. If you've ever listened to a fundamentalist radio program you'll hear them say, "The Bible is truthful." It's not necessarily factual, but it's truthful. So you have these distinctions. Now here I'm not going to play that kind of a game. I'll be perfectly outright and if it turns out that your presentation was faulty, but the facts were good, let the facts stand for themselves, let the arguments stand for themselves. If your presentation is good, but the facts are faulty, that doesn't count. So it's not the quality of the writing, it's the quality of looking into the facts that support your argument. Something that's contrary to what I impart. I come up with enough dogmatic opinions, that's no problem taking issue almost anytime with something.

ANN: Can we write something that is sort of an additional thing, that you didn't perhaps think of?

SASHA: Fine! But, something that would contribute to my writing, and I'll accept it.

It's going to fly right in the face of the rules of the department of chemistry, but that's going to be one of their problems.

[Directed to student] Yes.

STUDENT: You mentioned we have to take something that you state in the text. Could it be something that we remember that you said and disagree with?

SASHA: Sure! Absolutely, absolutely. If I in turn contradict myself, you can call me on the contradiction and straighten me out. That's fine. Because I'm apt to say one thing today and then in a fit of enthusiasm say something that appears to be contradictory. Tidy it up. Anything like that, that would be constructive to me and give evidence that you have somehow listened instead of taking notes.

So, generally, what about taking fewer notes and not worrying about it that much?

I want to get on, in the fleeting hour that's left, to the discussion of consciousness, which is more or less the title of today's talk. Consciousness, and maybe from there into mental illness. But it's a term that I don't quite know how to define. I don't think many people will define it with much consistency. I'm not quite sure what I can say about consciousness, except that a lot of the drugs that we'll be talking about are used because they give a change in consciousness.

[Directed to student] Yes.

STUDENT: Do you think the term "conscious awareness" may be a more apt term than consciousness?

SASHA: Conscious awareness? What's a change in conscious awareness? The term "change in consciousness" is common.

STUDENT: Consciousness is like you're conscious or you're unconscious.

SASHA: Well, conscious has subdivisions, the subconscious, the unconscious, different levels of psychiatric interpretation.

STUDENT: But aren't they somehow conscious awareness to some degree? I mean they might not be openly accessible to your normal state of conscious awareness.

SASHA: Okay, I'll accept that as being a usable synonym. Conscious of what? Of your surroundings in a way that you can recall and you can record.

ANN: Except that you can have unconscious memories.

STUDENT: In conscious awareness, conscious can mean deliberate awareness.

SASHA: Mmmhmm.

STUDENT: So if you take a drug that you don't know what its effect is and it affects your awareness, is that conscious awareness?

SASHA: When you're asleep and having a real rolling dream about falling down out of an airplane, is this conscious awareness?

STUDENT: Not deliberate awareness.

SASHA: Not deliberate.

STUDENT: But somehow recorded.

SASHA: It's somehow recorded and promptly forgotten.

STUDENT: Yeah, but it's going through some sort of—I mean it's like you have a photographic screen and things might fly by it, but might not be recorded on it.

SASHA: I think everything, almost everything that you're exposed to, in dream, in sleep, in falling asleep, in waking up, in a drug state, in the process of being born, in the process of dying, everything is recorded. I'm a firm believer in that record being there. You have so many instances in which you have the Proustian remembrance of things past in which you are doing something and a smell comes to the nose, or you're tasting something that triggers a complete gestalt of a memory of x years ago where you suddenly hear a certain sound or hear a certain note, or have a certain input and you have no knowledge of why that should trigger what it triggered. It does unfold an entire area.

There is a device that is sometimes used in surgery when you have the head open; when psychologists play alongside neurosurgeons. A lot of surgery with the head open, where you're actually going in there and thrashing around and taking out something that shouldn't be there or that was growing a little too large, is not painful surgery; it does not require total anesthesia, and you often have the person vocal. They can't see that part of the skull is laid open and the surgeon is fishing around inside of there. There's no pain; they don't feel it. So deep anesthesia is not necessary for brain surgery. You want topical anesthesia so they don't hear or feel the clipping of the skin and the drilling of the bone. You're in there and there is not much that needs anesthetizing. You'll get into the temporal lobes, the lobes coming along the sides of the brain, and there is a lot of feeling that these lobes contain memories. The whole concept of what memory is is really a very uncomfortable thing in the realms of people who love prime numbers and chemical formulae because there is no way of saying what a memory is. Is a memory a re-synthesis of proteins into a new form? But proteins are all the time changing. Is it an arrangement of neurons into a network that has a unique property of carrying that memory? Well, how do you possibly consider neurons hooked up in a way that you can remember someone's middle name is Jonathan? I mean, it's hard to envisage that kind of memory. And yet, it's in there.

Psychologists will tap little aspects, called "coordinate areas," in the temporal lobe and talk to the person who's under surgery. Tapping one, the patient might say, "I remember a time when I was eight years old and I was running down the sidewalk outside my grandmother's house and they had just painted the picket fence white. I had a stick in my hand, and I was running down the sidewalk and it was going r-r-r-r-r right down the fence and I had not realized the fence had been painted recently until I saw I had left a black mark. As I looked back on that black mark I saw it wandered up and down and I wondered how, if I was running straight, I could make such a wiggly mark." They haven't recalled that from the age of eight in forty years. Touching these areas can awaken memories that have not been accessible to the person for decades and suddenly the whole card deck is in their hands and they recall a whole incident. It's in there. You don't know how to touch it, you don't know how to get at it. You have things that are in there, like a little gap in a computer disk when the head bounces on the disk and there's suddenly a big area that's opened up and nothing is there. Aphasia, in which you have a complete knowledge of what you wish to say except one word, which is this word in there, has gone completely out the window. And you have no idea, maybe a person's name, or maybe the declension of a verb, or the choice of a word. In some ways, you can see a bit of this in your own experience where you're looking for a word, you can visualize it, it's about this long, it has an "e" in the middle of it and maybe ends with an "eng." You kind of visualize it, but you can't haul it out. Sometimes when you haul it out of context, you sort of look to the side, you can get it. There are all kinds of tricks, none of them dependable, for recalling these kinds of things.

But sometimes that memory is just totally wiped out. You realize what you're looking for was the name Charlemagne, because that was the word and there's no way around that one word, and even when you hear the word, it's totally strange. You could memorize it, but you're memorizing it for the first time. Something has gotten in there, in Korsakoff Syndrome, then another word pops out. You don't realize these names and words and dates are disappearing, until pretty soon you're trying to put a fabric together and you find it's got holes in it and you don't know what's in the holes. And when the holes are filled by someone, they're not familiar. It's a form of playing with memory.

The whole area of memory which is part of consciousness, or conscious awareness, to tie these areas together, has been studied by trying to determine how long it takes to get what they call short-term memory into long-term memory.

For example, when you were driving to work or bicycling to school this morning, a car cut in front of you and you swore at the car and for some reason you noticed the license number was 124GTN. Four months later, 124GTN is so far out of recall you can't haul it out. That's short-term memory, the thing you want to know right now. How much money have I got on my BART ticket, do I have to stop afterwards and refill the money? This is short-term memory. If you could recall the amount of money on every BART ticket you had handled for the last six years, you would be in an absolutely raving heebie-geebie memory state. Amnesia is a life saver. If you could recall every word and every act and every dream, god think of that—when we get into the term schizophrenia, you're into something very close to this. That short-term memory has a virtue. Things disappear.

But long-term memory is where you have something that you can recall later and later and later, it's back in there if you know how to get at the file, so to speak. And I'm not quite sure there's a short-term that goes up to here, and a long-term that goes there, it's sort of a continual level of reinforcement of importance. You want to paraphrase a book you've read to someone, and you want to paraphrase it with a certain amount of emphasis, you'll sometimes look in a mirror and you'll talk to the mirror and try to get your conversation together as if you were running it through ahead of time. What you're really doing is committing it to a slightly longer-term memory. But we have, I think, a memory that really embodies the license plate you saw on the car that cut in front of you on January 3, 1984, and the amount of money that was on your BART ticket, and how your name had been misspelled on the letter you received from someone four years ago. I think it's in there. It has no survival value as such, as far as you know, until sometime out of what's called the unconscious comes this welling of a response to something that is not learned, is not thought through. You'd like to call it instinct, yet we look upon ourselves as intelligent animals. Somewhere we have made a choice based on what has been put in our memory previously.

So, I want to get more or less through the hour today talking about consciousness, conscious awareness, conscious state, state of consciousness, change of consciousness, without getting into drugs. What goes on in there, what it's like in there without the drugs. We're talking about drug altered changes probably for the rest of the year, so I want to spend some time to give emphasis to the fact that drugs don't do these things: drugs allow these things to happen, sometimes with more ease, sometimes in different ways. How many people have dealt somehow with the concept of a catalyst? Something that either allows something to take a different course, or that changes the rate in the course it would otherwise take. But something that permits an event that is going to occur to occur with more ease. Usually it's an event that would otherwise occur anyway. If the change of consciousness is a desirable thing, a drug can achieve that.

I think there's an argument from Andy Weil, not so much in this book as in a previous book, *Natural Mind,* in which he holds with the thesis that the impulse of the human animal is to seek out changes in consciousness. It seems to be a built-in something in the person. And you look around and you see that. I mean, golly, go into a good rolling saloon about 10:30 at night and you'll find it really broadcasts altered states of consciousness. And people have not been told, "Go in the saloon and drink or otherwise we'll take away your social security card." No, it's a direction that is gone after. People will choose to go down roller coasters at 10,000 miles an hour and then go up the other side going "Eeeeh! Aaaah!" What in the world do you want to frighten yourself to death for? [Laughter.] Why do people stand in line to go on a bumpy ride? It alters something inside. Kids, as Andy Weil has observed in his own medical practice, will sometimes give themselves the equivalent of the Heimlich maneuver, they'll grab their chest, they'll breathe real fast, in and out, in and out, take a big breath and squeeze themselves like that and virtually pass out. How many people have done that? [Laughter.] Ha!

Probably you'll find many, many classical allusions to getting at the human mind, which is really a remarkable structure, and its awareness of what's around, and its awareness of its own illness, its own pains, and its own limitations, and how it is at peace with that. I think a good starting point is a person who is probably not known at all today, but it's a nice name to know. How many people have heard of Anton Mesmer? One, two. Okay. Three. Mesmerization, mesmerize, the term to mesmerize. Not hypnotize, mesmerize.

I remember the very first time I really achieved a successful mesmerization on something was with a chicken. I lived on a farm for many years and what I'd do was toss a piece of corn in front of a chicken. And the chicken would look around with her one-eye routine and spot the corn, and as soon as she put both eyes on the corn, neither eye was on the corn. But her beak was pointed in the right direction. And as the chicken would go down toward the corn, I would touch the corn with a piece of wood, just a fine pointed stick, and as the chicken was just about to make that lunge, as she was seeing the thing from both sides, I pulled the stick away and the chicken would stop in mid-lunge. Then I would pick the chicken up, move the chicken, put the chicken down, put the chicken on her side, put her straight up, no response! Whatever was going on was completely blanko inside there. Then I would put the chicken down in front of the corn, take the piece of wood and push it back to the corn, the chicken would eat the corn and go look for another piece of corn. A good example of mesmerization.

This was first developed by a person who was a true magician in his own way, a person who in this age probably wouldn't survive. He didn't even survive in his own age. He was in the wrong time. There probably has never been a time for a person like Mesmer. In France, before the big revolution in the late 1700s, Mesmer had taken a study in law, then he got into divinity, and he finally got into medicine and took his degree in medicine. He was a very firm advocate of a theory of the time, that was just sort of building, that illness was personified by pain. In fact, pain was the representation of illness. Now we have no idea of the role and the fear and the dread and the despair that came from pain then, because now we have a whole family of drugs that are geared toward relieving pain: analgesics. But at that time there were very few drugs. You could stupefy, you could give a person opium until they fell over unconscious and they would not remember the pain. But pain in the conscious living person is a dreadful thing when it cannot be relieved, and there is means to address it.

It was believed at that time, and Mesmer was one of the advocates of the belief, that there was a vital fluid that ran from the stars to the earth to the people. This fluid was a kind of a circulating, moving thing that was in dynamic balance. And pain was an imbalance of this fluid. And if you could only get the fluid rearranged, get it back into its proper channels again, you would be free of pain. The term that Mesmer used for this concept was "animal magnetism." Because at that time we were just getting to the Age of Reason in which magnets were found to do remarkable things and no one could explain them since they were magic. And he would get a couple of magnets and go up to a person in pain and align the magnetic field, he'd align the fluid flow of that person. And the person would find the pain would go. And, by golly, there are two people knocking at his door, and two people would get his medical treatment, pretty soon there are four knocking at the door. And more and more people heard of Mesmer and his ability to realign animal magnetism. And of course everyone paid him, and the cream of the cream came around because people who were in wealthy situations or in the court had pains that were just as devastating as people who were in poverty and working in the fields.

And he finally couldn't handle the large crowds that would come. So he invested the magnetism into other things so he could use them. The first thing he used was a long piece of wood. He magnetized the stick with his magnets so he would merely need to point the stick at people.

He would sweep through a room, and the pain would disappear. This is the beginning of the magician's wand back in the late 1700s. Pretty soon he found he didn't need the stick. He could point, he could touch or even point at people and the pain would disappear. Then he invested this more deeply and found that he could magnetize a mirror and the person could look in the mirror and relieve their own pain. He could magnetize a harpsichord, and when someone played the harpsichord the notes would relieve the pain when people listened to them. He would magnetize a forest and when you were tied to a tree in the forest, you were caught in this magnetic field and it would relieve the pain.

I don't know to what extent it was really his magic or to what extent he had found a vehicle into a person's own awareness and their own capability of healing themselves. I rather suspect the latter and I rather suspect that you suspect the latter. But it is a fabulous story. It is the first example of the relief of pain without going in there with a glob of something out of the opium poppy. Or eating something that came out of the marijuana plant. Or *Mandragora,* the mandrake root. Belladonna was known at the time. These were about the only major tools available and they only relieved the pain in that you didn't remember the pain when you were under the influence of the drug, but you did remember the pain afterwards. None of these drugs had lasting relief. But his methods had lasting relief.

The medical community was totally skeptical, but it couldn't be too skeptical because he was a physician. The French government said, "We'd like to set up an Institute of Animal Magnetism. We'll put you in charge, give you staff and resources. France will be the leading country in the world in the treatment of pain and the treating of illness in this way." He said, "That's an absolutely lovely idea. I'll go along with it. There will be an institute and I'll certainly teach others. After all, magnetism doesn't last and has to be continually reinforced. And I won't live forever so I want to teach other people how to do this. There's only one thing I want. I want to get recognition from the Academy of Science that this is a valid approach to the practice of medicine."

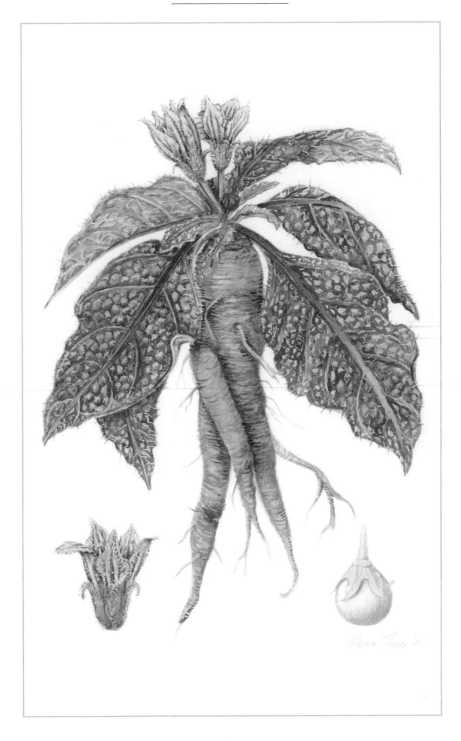

Well, the Academy of Science was made up of a lot of people. Ben Franklin was a member, as a matter of fact, because he had done work with electricity and magnetism was close to this. And the general consensus was, "We don't really believe it is valid because it smacks of charlatanism and it probably has a bit of Satanism wrapped into it. We certainly can't explain it. No, we are not going to acknowledge it in the Academy of Science." So, there was no institute. He could not get peer recognition for what was a dramatically, easily demonstrated skill. And about that time the French Revolution occurred and everyone who was his customer, the higher court and the people of affluence, were beheaded. It was rather a total change of consciousness in many people's hearts, so he scooted from France and ended up in Vienna where, since he came from the area of the revolution, they assumed he was probably part of the trouble, so they put him in jail. So that was more or less the end of Mesmer's exploring.

But some of his students were exploring more the "enchanted forest" argument. They would tie a patient who was in pain to a tree and then also point to the person, so they received a double whammy of the "enchanted forest" and the "enchanted hand," with the result that the person went to sleep, or appeared to go to sleep. They became somnambulant; walking in sleep. When they untied the person and let that person be free, the person would do what they said: move around, respond as if they were voluntarily doing things. Then they would tell the person to wake up and the person would not have any recall of what had gone on. The first touchings of hypnotism. Originally it was called "neurohypnotism" and then became just "hypnotism." Again, totally contrary to both science and religion, and also law.

Hypnotism has probably been known over the millennia in different cultures at different times and been lost. It was rediscovered in this country in about the mid 1800s in which one person was studying a person undergoing hypnotism and realized, just as the person went into this trance (trance is another area that is not much medically recognized), the eyelids would flutter. And the hypnotist realized there was a concentration factor at play, and began saying "Concentrate on this, concentrate on that." You know it as a stage device where you put a little object wiggling back and forth, "Look at this and you're going to go to sleep." It's almost a stage game, yet it is really an extraordinarily powerful medical phenomenon that is probably now being pretty much recognized in medicine.

For decades in this country, the medical community said, "It is a magic parlor trick and has nothing to do with the practice of medicine." And yet, about 90 percent of people can be hypnotized in the sense that they lose awareness of where they are, enter a dreamlike state, and yet remain responsive and physically sufficiently coordinated to respond to instructions and to orders. Of that 90 percent, probably only 10 percent are deeply enough hypnotized to allow painful surgery to occur. It is not that good a method for surgery. But it is, in 10 percent of the cases, quite adequate for otherwise painful surgical procedures. The pain is not remembered. You can never tell if pain is not felt. There are instances in medical hypnosis in which a person will scream with pain during certain interventions and afterwards say, "Thank goodness for the way you did that. I felt no pain." This pain that is not remembered, but felt, becomes an interesting point of philosophical game balance.

Has anyone here worked with hypnotism? It's not in my territory.

STUDENT: I think there has to be a willingness from the person who's going to be hypnotized. I mean, they have to want to be hypnotized, you know?

SASHA: I've heard that.

STUDENT: Some people who have this control syndrome, you know, "I've gotta be in control," really have a hard time letting go.

SASHA: And yet I know of demonstrations in which there has been a rather effective mass hypnosis of a group in which not everyone is a believer and/or particularly consenting. So I don't know, I don't have the answer. I don't know how one would determine the answer.

[Directed to student] Yes.

STUDENT: It seems that a lot of sessions relieve pain. Is this an old—?

SASHA: From the very, very beginning of humanity everything was invested with soul, with personality. The rocks had personality, trees had personality, animals had personality, foods had their own soul. And it was this conflict of souls that led to illness and led to pain. Pain was an expression of this conflict in the very earliest of primitive peoples. And the way you relieved pain was to resolve the conflict. Hence, beating the demon out of the person, the demon that was causing the pain. You'll still find that in some places today. You'll find that insanity, like schizophrenia, has been treated by taking the person and beating the dickens out of them with a stick.

Interestingly, very early in the structure of human development, the shaman, the witch doctor, who controlled illness was often the man in the group, but it was often the woman who was invested with the magic for pain control. Because a woman brought with her as part of her heritage the birth of children, which is a painful process. Therefore, she was in the position of knowing the structure of pain, knowing the positives and negatives of life having lived with it and survived it, so the power was vested in that person. So very often it was the woman who was the shaman of pain, whereas the man was the shaman of other demons. That's a whole structure; I don't know how to really get into it.

I've talked quite a bit about learning so that you have a freedom of choice, to choose what you want to do. If you know what a drug will do, you can choose to use the drug or choose not to use the drug, but do it from a position of knowledge.

There is also another very, very subtle thing that you can give up besides choice, and that is to give up your own power to do things. For example, if you are ill and you say to someone else, "You know what's wrong with me. You cure me," you are in essence giving up your power to the other person. And the other person is really working with potentially no more than you have, but you choose not to work with what you have and to work through someone else. You're either giving up your choice to someone else or giving up your power to someone else. Both are the losses of very, very effective aspects of your personality. We'll find structures of this coming in when we get into some of the current day arguments on drug abuse testing.

How many people got the handout that had the definitions? Did everyone get that? Has anyone read through it to the hallucination area, the definitions for hallucination and imagery and such? Good. I want to talk a little bit about this, because when you talk about change of consciousness, you've got to realize where your consciousness is so you can recognize what the change is. No, I am going to leave the topic of hallucination alone as I am going to take another diversion. I want to talk about what is commonly called brainwashing. This is, again, a change of consciousness and it has been a misapplied one. The association, in most people's minds, with brainwashing is with Pavlov.

Pavlov was a physician and psychologist in Leningrad at the time of World War I whose name is associated with brainwashing, the "Pavlovian response." People remember him, they remember the "Skinner box" as a given action, a given reward, and pretty soon the reward itself is enough to trigger the action. You take a dog, you ring a bell, you give the dog something to eat, the dog salivates. Pretty soon you ring the bell and the dog salivates because the connection has been made between the bell ringing and eventual food. But the food is not necessary. The bell serves as the instigation for the salivation.

This is conditioned response. This is perfectly straightforward. We are all very much happier in our lives in many ways because we have allowed ourselves to be trained to conditionally respond to various stimuli. It keeps us from having to think. It keeps us from having to make decisions very often when decision time could be used better for other purposes. Don't ever forget that the decisions still have to be made. This is an issue I may have brought up, the idea of doing things by habit, where doing things by habit robs you of the freedom of choice. You tend always to open the door with your right hand because you don't remember that your left hand can do the job too. But you've gotten into that habit because you've gotten into the habit of carrying your books under the arm of your left hand. When you do something by habit, you've lost your freedom of choice. So bear in mind that little twist of the freedom of choice.

The conditioned response is a way of getting you into habits. The real thing that Pavlov discovered was totally by accident. It was roughly around 1920 or 1921. There was an unprecedented flood in Leningrad. Leningrad is a city of canals and the water level is slightly higher than Venice. It is a city in which floods can be quite devastating. And just after World War I, there was a flood in Leningrad and all of Pavlov's dogs were in their cages in the basement of his laboratory and the basement flooded. What he found to his dismay, on finding the results the following day, was about nine-tenths of the dogs had drowned and were dead. What he found to his amazement was the one-tenth of the dogs that had survived with their muzzles being right up against the top of the cages, getting what little air there was, had to a large measure reversed their conditioned responses. That approach to the moment of death, the certainty of that death, and then the relief of the death not coming, changed their whole attitude. Those dogs that salivated when they heard the bell now no longer salivated when they heard the bell. They bit whoever was there when they heard the bell, or something totally different than before. There is your brainwashing.

How many people have read books on the near-death experience? NDE. Quite a remarkable situation in which you realize you're going into death, and reports have been varied in their character, but the direction is more same than not. It is very culturally dependent. In India with the near-death experience, when people see things, it is not as a bright light at the end of a tunnel with a feeling of being at peace, but of seeing magistrates with big charts up in the pigeon-holes and coming down to fill out the final chart. The culture will dictate the structure. But you are going into an area in which you can see yourself, very often from a distance. There have been many instances in which people reported in surgery to have observed the surgical field from god-knows-where, but being able to see that poor body down there being worked on under anesthesia, feeling no pain. Again, you're in a depersonalization, you're out of the body.

I was talking to one person who almost died as an infant in childbirth. And she remembers vividly, from the unconscious, in a very dramatic moment the whole thing came back to her, her memory of being born. She found herself looking upon her mother bearing her, from she'd like to say above, but she had no real knowledge that it was above because it wasn't from anywhere specifically. But she could see the entire scene and was realizing that this child may not live. And the child did live, and it was she.

So that kind of phenomenon, where there is actually a true change of state of consciousness. The idea of approaching this death thing with the reverberation of a whole philosophy, again and again and again being reinforced. You see that person die and this person die, and another person die, and it is descending upon you and you don't know if it struck or not, and when you suddenly realize that it did not strike, then that philosophy has been imprinted in some extraordinary way, totally made part of your fiber. This is the brainwash. This is the conversion process where you suddenly take up a whole new attitude toward something. It could be a delusion. You'll find illnesses that are defined in this way. The delusional process, the paranoid delusion where something conspires to make a slightly different view of something that is reinforced and built-in, until it becomes totally part of your fabric.

I had a phone call about four years ago from a person I knew moderately well, not a friend, but a person I knew fairly well. And he knew me and trusted me enough to phone me. The first thing he said on the phone was, "They're out to get me. They're gonna get me. And they are getting close." "What's going on, Jeff?" "Don't give me any condolence. They are really out to get me." He's a psychiatrist, MD, in research, and something conspired to give him the impression that something that occurred was a way of undermining him, and everything else he saw either reinforced it, or if it didn't reinforce it, he found fault with its being valid. True paranoid delusion. I think everyone has had a touch of this. And probably everyone has known someone who has really been caught in the throes of a paranoid delusion.

I said, "Jeff, who's out to get you?" "Oh, no! I don't think there's anyone out there I can trust." "Well, have you considered seeing a good physician?" "They're the worst of them! They've actually got my father and they gave him brain surgery to convince him that I could not—" this and this and this. The story just got monstrous. And I was in an interesting situation. He could not talk to anyone, but for some reason I was at a sufficient distance and was of a sufficient age over him that somehow, he felt he could trust me. Think of yourself in the odd situation, if you are caught by a dear person who you know needs help and they're in a paranoid delusion and you realize no matter what happens it reinforces that delusion. There was nothing I could do to unhinge it.

If you say, "Come off it, there is nothing going on, forget it, they're not really out to get you. That guy wasn't really following you and he turned off at the other corner anyway," he knows very well he turned off at the corner, but he wouldn't know if that person was following him. You are trying to dissuade him from the validity of the delusion, so you are part of the conspiracy. If you say, "You are absolutely right, I've read the papers too and there is no question that the Republican Party has the whole FBI and CIA out after you, and it is really going to go the way you think it is," on some unconscious level he knows that this is not so, and he knows this is a paranoid delusion, he knows this is an illness. Now suddenly you are trying to reinforce his illness? You are not an ally at all. In such a situation you can't win. If you try to help, you're part of the problem. If you try to dissuade, you're part of the problem. This is the beauty of a real totally spun-in paranoid delusion. You are part of the problem whatever you do. A visitor walks up; they're part of the problem. Someone walks by on their way to church; they're part of the problem. All you can do is listen. Listen and don't try to volunteer suggestions. Listen and be there, support, and somewhere, somehow, get them to where you can break the feedback cycle.

In this case it was very nice, we took advantage of his father. His father was in some sort of potential psychological problem himself, so I said to him, "Jeff, your father's back in New York and he needs some medical psychological help and I don't know if they're going to treat him correctly or not. You are probably in a good position with your medical background to be there and try to oversee what they're going to do to your father." "Okay. Okay, that may be a good idea." At which point we alerted the people who were working with the father to watch that Jeff was coming in a really intense paranoid delusion, and to get him to the medication. It's Jeff who needed treatment, but the father was the foil. It got him there, got him to where he acknowledged that he was having some inability to quite see things in a logical way. The medication that's used in such cases does not treat the delusion, it breaks the feedback cycle that maintains the delusion. It can be done. Nature can take care of it. Something begins unravelling in the same way it raveled up. A good case of a mental illness that is very, very hard to treat because there is no anti-delusional pill you can give.

In other areas of mental illness, you have schizophrenia. We talked about that possibility if you could remember all of your dreams, if you could remember that dream from last December, or if you remembered the dream from the last day in September, imagine what your switchboard would be like upstairs if all this sort of thing was continuously coming into you and you just couldn't pick and choose. Let's say you had to cross a street and everything from the far right to the far left and from the sky to the ground was coming into you. There was a car coming from the right, the light was green, there was a bird circling over there, there were a number of insects coming along the line, a dog was coming down the sidewalk, everything that you can record was coming into you. How are you going to cross that street safely? You've got to be able to say, "I'm going to ignore that, I'm going to ignore that, that has no immediate importance, that and that are of importance. I will observe the light and the car, and kind of glance that way just in case." Three things that may be important. And you'll make it across the street, there is a pretty good chance, unless something falls from the sky. Admittedly, there may be an airplane in the sky dumping fuel and that's the very thing that got you. But if you are looking for the airplane and seeing it, you may be looking for the fuel that never comes. A schizophrenic state is like a switchboard in which everything is lit up, all connections want communication, all things come in and you don't have the ability to put things in priority. Very often a person in this sort of state doesn't have the sense of where they are. There's a term used in psychiatry: to have your center. It means the center of knowing where you are. Sometimes you'll talk to a person and you'll find that person is just aside from their center. They are not really, totally there.

I have a very firm belief that people who make the best actors are people who are never quite in their own center. And so they can go into the center of a character they play and they can be that fictional person and superbly represent that character. But when they're off the stage, they're again a little bit to the side of themselves, not quite their own person.

I had a girl in a class one time at the Institute of Asian Studies (I was giving a course in psychopharmacology) and I noticed that her throat was cut, there were big scars on her throat. One time she wore a short-sleeved blouse and there were scars on her arms. She was, I would say, a little bit to the side of her center, but she was very interested in the course and interested in psychopharmacology, and especially interested when I got into the area of the heavy tranquilizers, the antipsychotics, phenothiazine type things. She was very fascinated. I got to talking with her afterwards, she said the phenothiazines really kind of make a zombie out of you. And indeed, they do.

How many people have taken chlorpromazine for motion sickness or for diarrhea or something? One, two, three, four. I once got into a situation with chlorpromazine from which, I assure you, I can tell you what a zombie is like. This was over at Langley Porter where we were pursuing what was called the "pink spot of schizophrenia." Every effort had been made for years to classify some way you could plug something into a person and read what comes out on a dial, or put something in them and if it comes out pink, they were schizophrenic. This one thing stemmed from a study done in Canada by a person named Abram Hoffer who found that if he took the urine of schizophrenic people and did this and this and this to it, made an isolate, put it on paper, run a chromatograph, spray it with something or other, he got a mauve spot at this location. If the person was not schizophrenic, he did not get a mauve spot. He had discovered a chemical test for schizophrenia, he was going to be famous. He, in fact, called the presence of the mauve spot "malvaria," which I think was really a sign of inflation to actually give a name to the thing he observed to classify the diagnosis he had built up. And he was promoting this as a way of diagnosing people being schizophrenic or not.

Well, I was working at the time at Langley Porter. The setting was kind of neat, so we thought since we had schizophrenics coming in all the time, we got them while they were florid, we'll take their urine and run the test. We'll take the next ten that come in, and we'll take the next people who come in with broken arms as controls. Hopefully a few of them will be schizophrenics and a few of the schizophrenics will have broken arms and we'll get two populations. We'll run their urines, we'll see how many are mauve positive, how many are mauve negative. We ran twenty urines. All mauve negative. What's going on here? We got hold of Hoffer on the phone, "How did you do that? We followed the directions." "We've changed the directions." "Give us the new directions." We took the next twenty patients and followed the new directions: all mauve negative.

We got back to Hoffer and spoke with his technicians, "Well, we're just doing what we say we're doing, and we're getting 89 percent validity and mauve positives." What's going on here? I worked at the time with a fellow named George Elm. We were talking about this over a cup of coffee and suddenly, bingo! Where is Hoffer getting his schizophrenics from? "We're getting them walking in off the street as new admissions." He was working at the university hospital at Saskatoon.

"I wonder if, I wonder if—" we said. And so each of us took 250 milligrams of chlorpromazine.

It was my first experience of becoming a zombie. For about thirty-six hours it was a strange place. Not much memory out of it, depressed, unable to move around, you don't care, you are totally disconnected from your emotions, from affect, from feeling, from what's going on inside, *yeeaach*! I don't know how else to describe it. Pretty soon, about 30 hours down the line, you see a little bit of light out from this "yee-aach!" You begin climbing out and pretty soon you're back in the real world again. But for thirty hours we were both mauve positive. It turned out that Hoffer was using schizophrenics in the hospital who were being treated with medication, and he was picking up the medication as being the metabolite that was giving the mauve color analysis in the urine. This was an interesting experience that taught me what being a zombie is like and, believe me, it's an altered state. It's not one you choose to go into, but know your altered states so you choose this one and don't choose that one. I would not choose chlorpromazine as a turn-on for the weekend. [Laughter.]

The point I'm making is there is a tremendous drive to find some biochemical marker, some biological marker, some test, that will say this person is insane, that person is not insane, this person is mentally ill, that person is not mentally ill, so we can classify easily by an objective test and put these in the hospital and treat them, and leave these out running around doing their thing outside the hospital. But this has been almost a total disaster because I really don't believe there is that kind of a conspicuous biological abnormality going on up there, any more than you can give a chemical test to determine intelligence, happiness, sadness. You've got to communicate with a person.

Consider what we call mental illness here and go to another society. I had an interesting interaction with the head of the department of psychiatry at the University of Moscow about a year and a half ago in which we were talking about the practice of psychiatry and the use of drugs in psychiatry. He was a very open person, but he was totally oriented within the Russian philosophy of mental illness and psychiatry. You hear a lot of criticism in the United States about their practice of medicine, especially the practice of psychiatry. In Russia, if you don't agree with the state, you're obviously mentally ill and are put into the hospital for being mentally unsound. We say that's not the practice of medicine, that's not the definition of mental illness. But their attitude towards ours is that it's not the definition of mental illness. It's an attitude and how you define it, and theirs is from the point of view of a society that is oriented toward social structure. We're geared toward the individual, and they're geared toward social structure. That is the pattern that has been built into their society for three centuries. It is not a communist phenomenon; it's been there for the history of the Russian entity. It is the collectiveness. Collectiveness has very much been a part of Russian history. For example, a person who wants to emigrate is obviously not totally together. There is something totally strange in that person's makeup. Why would a person who is mentally sound want to leave? It doesn't fit. And we have a touch of it here too with people who want to go to Russia. We had about 30 or 40 people who wanted to go back to Russia, some weeks or months ago, and it was asked "What kind of traitors are those?" We used the word traitor. "They're leaving this country to go back to Russia? What's wrong with them?" That attitude is also in Russia, when they are leaving to go to Israel or Scandinavia. "What's wrong with them?" That is kind of the definition of mental illness; it varies.

If you go into other cultures and other tribes, you'll find the definition of mental illness drifts around in a totally unsupportable way. It's an attitude held by a minority that is not in keeping with our social structure, our taboos and our various ways of conducting our society. Do you really believe that there is a misplaced methyl group in the biochemistry of the people who are Aborigines in Australia that defines their mental illness? That's going to be the same as the misplaced methyl group in us that defines ours, in Russians to define theirs, in the Arab countries to define theirs? Every culture has its own definition of mental rightness and mental wrongness and there is no total universal syndrome that exists.

So the idea of finding a biochemical marker—I mentioned the methyl group because a lot of energy has been put into the methyl group. It's the simplest group in chemistry, a one carbon situation. Building molecules is like playing the piano. You can hit one note or two notes or five notes, but you can't get anything in between. It's this or this or this note. Chemistry has got one atom or two atoms or seven atoms or twenty three atoms, but you can't build a molecule with one and three-quarters or one point seven two atoms. So you go to the simplest possible group, the methyl group, a one carbon system. That group has invested in it more man hours in efforts to find its role in mental illness.

And there are some interesting findings. In a handout about three lectures ago I gave a picture of methionine, an amino acid, the animal amino acid, the trap to people who are total and devoted vegetarians. They'll find that they will run into certain nutritional deficiencies because we've got flat molars. We're omnivores. We are geared in our evolutionary process to chomp on meat and chomp on leaves and eat eggs now and then, and multiple nuts when they're available. This is the structure that we have evolved with. And so total vegetarianism is fine for a philosophy, and if it cures you of illness, fantastic! But, there are certain needs of the body, and methionine is a superb amino acid that is one, that comes largely from animal sources. Methionine is called an essential amino acid because we don't make it. So we've got to get it from outside. Methionine's role in the body is to transfer a methyl group from here to there, its methyl group to that over yonder. All of our neurotransmitters with possibly one exception, but certainly serotonin, norepinephrine, epinephrine, and dopamine, all require methyl groups to become inactive. The body gets rid of them by, amongst other ways, by putting on methyl groups.

Almost all the things that are psychoactive, psychedelic, or things that cause a change of state of consciousness, contain methyl groups, and without the methyl groups they're not active. So the methyl groups are kind of spun into this altered state thing, by observation of what is there that causes an altered state. A methyl group is spun into the body as a way of inactivating things that are normally there. And what's fascinating is that some of the things that are inactivated in the body are really first-hand kissing cousins of things that cause altered states when they're taken from a plant source. So there is this great appeal: is there something in the body that has a metabolic process that goes wonky, that gets into a strange metabolic misadventure and the body plunks a methyl group onto something in an attempt to get rid of it and really turns it into something that causes mental illness?

Take the pineal gland, the little thing that hangs down in the middle of the brain. The pineal gland takes something like serotonin and methylates it, puts an acetyl group on it, and you have a material called melatonin, the so called hormone of the pineal. This melatonin, which goes up at night and drops in the middle of the day, fluctuates on the circadian cycle. If you take this material, a simple hormone, take the pure chemical and put it in a beaker, put the pH at physiological pH, put the salt concentration at physiological salt concentration so that you are imitating the blood system, and stir it in that beaker, it will cyclize to form a new compound that is a derivative of harmaline which some consider to be an effective psychedelic.

So you can take a hormone and treat it under physiological conditions to convert it into a material that will cause an altered state of consciousness. My god, we've found the tool! Well, when you look at the pineal, it is invested (with centuries of lore anyway) as being involved in this whole territory: the third eye, the seeing eye, the center of the soul. It produces an alkaloid, a base, that under physiological conditions can metabolize to something that will make you a little bit funny. What an ideal solution! Obviously, something in the body causes this pineal, for reasons of misadventure or stress or whatever, maybe a bug or a virus, to go through this transformation, and there is the source of our mental illness. An extraordinarily appealing hypothesis. So close... completely wiped out.

One person I knew in Texas named McIsaac spent the last five years of his life going to auto accidents where people who had been schizophrenic had been killed, getting the pineal from the autopsy, and trying to find this chemical. There is no evidence that chemical is there in normals or in schizophrenics, or in paranoid psychotics or whatever form of mental illness you choose to look at. McIsaac was so invested in this, and became so deeply depressed by the fact that he could not tie the pineal, its hormone, and the known psychedelic that could be made from it into an explanation for mental illness, that he killed himself. He was actually the director of the Texas Research Institute for Mental Sciences at the University of Texas. There's an example of an altered state. He actually acted out his depression.

There is often a lot of investment in people's theories. I beseech you, if you get involved in the area of science, do not fall in love with the hypothesis, do not fall in love with your theory. Go back to the very excellent concept of Francis Bacon: try your best to disprove the hypothesis you have. Try your best to find why it's wrong. If you can do that one experiment that shows your whole hypothesis is built on sand, and you learn that it falls apart and is not worth a thing, say hooray! Then you can pick up what's left, gather it together, and build another hypothesis that might be closer to the truth. Certainly, the one you had was not very good. Do not try to prove a hypothesis. Do not design an experiment that will prove your hypothesis, because it cannot be designed. No matter how you design it, no matter how successful your experiments are, it only takes one failure to throw the whole hypothesis out.

That's close enough to the hour. Next hour I want to get into research and how research is conducted in this area, and then we'll get into the area of drugs.

February 24, 1987

Research

SASHA: To a large measure try to follow what I am saying, write things down if you need to, and don't hesitate to get in the way and ask me to go back over something if I get inspired and get moving too fast and forget to slow down between sentences, as I do tend to do.

Today I would like to talk about research. This is the last of the preliminary lectures before getting into what I really want to talk about for the whole semester, which will be about drugs, what they do, and what they are like. Research is a difficult concept to come to peace with. Generally, it's defined as a systematic search in a given area for facts and principles. But in truth, it's very much allied with the concept of learning, of getting new information. People tend to look at half of research and will come up to me and say, "Gee, I really like the area you're doing research in. I would like to do research in that area, too." Well, that's kind of neat, but, you know, you can't just go out and, as they say, "do research." How many people use that term "do research" as if you were to pick up something and arrange it in a certain way and, voilà, there's research, it's done, I'll photograph it, I'll send it in and maybe someone will applaud it. Research is not just searching for the unknown. It's often *not* searching for the unknown. You're counting the number of scales on a lizard's leg. A lot of people consider that to be fundamental research in reptile biology. But the number of scales is there. You're merely counting something that's already there. You just don't happen to know what the number is.

So, research is really asking a question. The question you ask very much dictates how you go about finding an answer. You can ask these marvelously global questions: What is the meaning of humanity? Is there a God? These are neat questions, and they can go on for a whole lifetime.

You're setting up a straw man for yourself that you can't answer. You might be the one who stumbles into the meaning of humanity and that's nice. But then someone else will ask the same question and not fully accept your answer, so the idea of that kind of a question is nice but it is not an easy one to answer. Is a person under the influence of a drug? That question is right up there with the God thing as there is no way, as far as I know, of answering that question either. There's no way I know of approaching that question and saying, "Yes, the person is under the influence," or "No, they're not."

You had a cup of coffee this morning. Are you under the influence of a cup of coffee? I can't find coffee in you. I could measure you and your blood, but I could never satisfactorily prove that you were under the influence of coffee. Or even of caffeine. Maybe I can find some caffeine. But is having caffeine in you being under the influence of coffee? I don't know. So a very keen point in research is to have a question that is a carefully asked question.

I teach a course in criminalistics about once a year over at Berkeley. These are people who are going to go and become criminalists. It used to be called criminology, but the word criminology has become very dirty for a number of reasons. Criminology is the whole art and study of the law, crime, wardens, prisons, probation, sociology, the whole shmear. That's criminology. There used to be a department of criminology at Berkeley and it was exactly all of this. What happened was they had a big thing at Berkeley about ten or fifteen years ago where students climbed on cars and held up placards and threw things at police. In this parade of rebellion, a couple of professors were carrying signs and happened to be professors who had tenure in the department of criminology, which is kind of a no-no from the point of view of the administration looking out of the third floor of the administration building seeing professors walk by. Get rid of the professors! Well, they've got tenure. You can't get rid of professors who have tenure unless they really do horrible things. So they dissolved the department of criminology. They just tore it up. There was no more school of criminology. They took part of it and put it into public health, they put part of it into sociology, part of it into philosophy, and the rest into some other departments. And amongst them they gave different names. In the department of public health they had a set of criminalistics. Criminalists are people who sit in a laboratory and look through microscopes and make slides and look at fingerprints, fire guns and collect bullets, and compare all kinds of things to find evidence that is presumably being used in court in the search for justice in the case of a person being charged with a crime.

Very often people who work in crime labs don't pursue evidence for the sake of justice. They pursue evidence to help the prosecution, which is not necessarily the same thing. When a person has been charged with such and such, there is a fundamental assumption of innocence that has to be maintained. If the assumption of innocence is not conspicuously maintained, it has to at least be maintained in your heart to perform a fair evaluation of the evidence. A person who is accused of something has the virtue, the prerogative, of saying, "I'm innocent. You have accused me. Demonstrate it." And so, the function of a criminalist would be to find the evidence that could be used in court to weigh the evidence for and against, allowing a judge or a tribunal or a jury come to a balanced conclusion on the basis of what has been presented. So a criminalist is the one who goes in and looks at the fingerprints, and, in the case of drugs, looks for drugs.

So a person comes in to a forensic lab with a white powder and says, "I seized this guy who was doing thirty-two miles per hour in a twenty-five mile per hour zone. I had suspicion because the windows were rolled up and there was smoke in the car. I saw the ashtray. I had enough evidence to presume to enter, so I entered, and I found this in the trunk. It's a white powder. Is it cocaine? Shall we bust him?"

Well, often what they'll do is they'll come into the crime lab and they'll say, "Here's a powder. What is it?" It's an example of the kind of question that you cannot tolerate, because you may never know what it is. It could be something that had been extracted from the toenail of a llama in 1912, and the llama's long dead, and the toenail is not to be found, and it's one of a kind. You're not going to find out what it is! It may be a lifetime task. You may find something coming out of a tree somewhere, beautiful sap. What is it? Well, it's sap from a tree, but I mean, what is it that makes it taste good? You may never know. Avoid the question of "what is it." Insist upon the question, "Is it cocaine?"

So the question going into research is a very important part of research. Ask a question generally such as, "What is the melting point of mescaline sulfate?" "Is it at that melting point?" "Therefore, is this mescaline sulfate?" "Can we charge this person with the possession of mescaline sulfate?" "Have you already charged them?" "Well, we have a couple of loose charges and we haven't filled in the blanks yet." That's not the presumption of innocence. That's presumption of guilt, we just don't know guilt of what yet.

So this is a whole philosophy that I want to instill very heavily there, but I at least want to expose here: what is your research question? In the area of research, this whole idea of searching out an answer to a question, getting an answer, finding out if this is cocaine, if this is a drug that is a scheduled drug, if this is a cause of an accident because we found some of it in the person's blood and they were driving down the wrong side of the street. I don't know, maybe they had gotten up quickly and were dizzy, maybe a bee was in the car, I don't know why they were driving erratically. Was there a drug there? Was the drug responsible? You have to answer these sorts of questions. But this is half of research.

The other half is often ignored, and that is the whole approach to the question, "is the answer already known?" What is out there that's known? I asked: what is the melting point of mescaline sulfate? Well, you don't have to go out and get mescaline sulfate and run its melting point. It may very well be in the literature. Some people devote themselves totally to that half and become library buffs. They accumulate information: "I'm going to really do something novel once I know everything that's known in that area." The other half are over here: "I don't want to know what's known. It will prejudice me. It will bias me. It will rob me of seeing things except as other people have seen them. I want to be totally *de novo,* creative and on my own. The heck with what's already known." Both marvelous, laudable positions, but the two have to be brought together in some sort of a sympathy in your own work and research.

So a goodly amount of research is determining what is already known. I had a student, not my student, but he wanted to work for me for a semester over at Berkeley on a master's thesis. And so, I listened to what was going on. He was working for someone else, but he wanted to use me as a sounding board. He said, "I'm going to work in criminalistics, and I want to determine how old the powder was that was in the bullet that fired the shell by gathering the fragments of the nitrated residues of diphenylamine that are there, that are a function of the age of the bullet. And I want to determine, from what products are there, how old the bullet is, so I can know if it's an old or new bullet." This is what's known as a criminalistic detail. They call it trivia and a lot of modern-day research is indeed carefully measured trivia. It's like counting the scales on a lizard's leg. The next leg may not have that number of scales. That's immaterial. You're doing a good job on this lizard [laughter] and you'll get a number. A lot of these things are sad excuses for research.

I subscribe to a publication called *Current Contents.* I think I brought one in today for another purpose. This is my way of keeping abreast of the published scientific literature in areas that interest me. Fantastic! Every week. This is one of a few volumes, this one is life sciences. There are also issues devoted to the physical sciences, the social and behavioral sciences, clinical medicine, engineering, and probably one or two more I'm not familiar with. Every paper that is published that week has a table of contents listed in here along with an abstraction of keywords, and the names and addresses of all authors. Every week. This shows the staggering volume of literature that floods the scientific libraries. And, I am sad to acknowledge, almost all of it is the equivalent of tallies of scales on lizards' legs or distribution of nitration products from an old bullet.

At random, let me open a page and we'll take that title: "Negative Reinforcing Properties of Naloxone in the Non-Dependent Rhesus Monkey: Influence on Reinforcing Properties of Codeine, Tilidine, Buprenorphin, and Pentazocine." So someone has a rhesus monkey and has a bunch of drugs on the shelf and they're desperate because they are coming up for tenure by the end of next year. I am guessing, of course. But also

"Too bad about old Ainsworth. Published and published, but perished all the same."

I will guess that researcher needs some seven papers a year in their bibliography to convince the faculty review board that they are a dynamic researcher. This is known in the academic world as "publish or perish." And "publish or perish" is a humorous thing, but there's a sad amount of it in the academic world. I saw a cartoon from the *New Yorker* posted in the Life Science Building in Berkeley. Two professors are walking across the campus, their hats on their heads and carrying their briefcases, one of them is saying, "It's a sad thing about old so-and-so. He published and published, but he perished anyway." [Laughter.] And so, there is a smell of that.

But there is this urge for getting in these trivial things which are not questions asked, they are answers found on the basis of what tools you have. Get a person a $100,000 instrument, who's been looking for this $100,000 instrument all their life. They go gung-ho over this $100,000 instrument. "What can I learn from this $100,000 instrument? Bring me your problems. I need five a year, to justify the support I'm going to get for the next five-year period." They become seduced. They become a prostitute to that instrument. They are not asking a basic question of, "What is the meaning of life?" or "Is there a God?" And this instrument might help give them the answer. They're caught looking for the inside-out down the other end of the telescope. This is the instrument which they want to observe the world through. And the world is not that observable. They're not asking a question.

So the question is not, "What drug is this?" The question is, "Is this cocaine?"

Back to my gunpowder student with his desire to find out the nitration. Here's a material that's added to gunpowder to make it last. You don't want free radicals floating around in gunpowder because it tends to get very brittle, very fragile, very sensitive. And so, as these little radicals are given up, you put in a scavenger that grabs them. Nitro is the giveaway of gunpowder. Something that is something-or-other-nitrate or nitro-something is added, to gather up the nitros. And so the longer the gunpowder is there, the more of these groups there are, the more complex these impurities would become, and sure, it's a neat idea.

I asked him, "What's known on the area?" "Well, I don't know if anyone's looked at it." "That's not the question. Not what you have found out that is known. What is known?" "Oh! Well, there are probably some people who have worked with gunpowder before." "Yeah, there may very well be. You have militaries around the world. You have industrial manufacturers that make their whole livelihood making gunpowder. Find out what's known." "Oh." Off he goes to a famous library and gets into the industrial and governmental literature.

He comes back in about three weeks and says, "Hey, there's quite a bit, but most of it is restricted information. The military has it and they're not publishing it." He was unable to get any details from either the powder manufacturer or the US Army. "Oh, well, what's known that's not been published?" "Well, not much. Apparently, there is some nitration." "Are the compounds known that you might make? You have diphenylamine with mononitrate, you have di, di, tri, tri, you put about sixteen compounds together, are these known?" "Well, I thought I'd make them up as reference standards [he had no knowledge of organic chemistry] and then just run up a TLC or chromatogram or something and see if they're present in the gunpowder." "How many of them are known?" "Well, I don't really know." "Go and find out."

He went out to the library, came back and said, "The *Chem Abstracts* only has seven of them." "What about the German literature from the nineteenth century?" "Oh, that's all in German." "Go out and look." He disappeared. In another three weeks he came back and said, "You know, it's amazing! They're all known compounds. Every one is known." "Well, we just saved asking a whole big question about whether the compounds are known. The answer is, yes, they're known. Are any of them found in gunpowder?" "Well, apparently quite a few of them have been studied and quite a few of them are in gunpowder." Almost every one had been synthesized and characterized about a hundred years ago, and his whole research project was essentially done. The question he asked was already answered.

So a good half of research, this other half which I'm going to talk about for at least half of this hour, is how you find out what is already known. If it's a drug, the approach is that you go to where there are listings of drugs. Let's take again the melting point of mescaline sulfate. There are books in the library. How many people have been in the library? Let's say you want to find out the melting point of mescaline sulfate because you have a white solid and someone says I think this is mescaline sulfate, why don't you take the melting point on it? Where are you going to find out? Looking in subject indexes under M? Not very likely. How about sulfate? Not very likely. Looking under melting point? Not likely, you are stuck. A goodly moiety, a good word from biology, a substantial chunk of research is finding out what is already known. Assuming it is about a drug, how many people have heard of the *Merck Index*? How many people have not heard of the *Merck Index*? Over half have not. This has been remedied. You have now all heard of the *Merck Index*. Do not confuse it with the *Merck Manual* which is a medical handbook for diagnosis. I'm talking about the *Merck Index*. The best buy you're ever going to make if you have a library and if indeed you're going to go on anywhere into anything that's called research.

Remember that you must begin accumulating your own body of things that you know, your own textbooks, your own reference books. One thing I love doing, I'm not very diligent, but I do it methodically, is to get one good fundamental book on every little discipline that I might want to go and get an answer out of someday. Even though I may never have read it, I know it's there. Begin accumulating a library. My office is surrounded by piles of books, but I know where they are. The *Merck Index*. A major one. The tenth edition, which is the current edition, is not much different from the ninth. So if you can get a ninth, get a ninth. It's probably going to be cheaper. But don't ignore the eight, the seventh, the sixth, back to the first edition. Every edition of *Merck Index* is different. The current *Merck Index* lists probably 10,000 or 15,000 drugs. It cross indexes drugs. It has empirical formulas. You don't know the name of the drug, but you know it has five carbons, nine hydrogens, a nitrogen, and two oxygens. Look it up under the empirical formula. Look up the atoms. Find out the compounds that correspond to that. It is a superb first reference.

It is also missing lots of things. It doesn't have plants. If you want to look up ipecac, you'll find the components of ipecac, but you won't find the plant. We mentioned a while ago Tuinal which is a mixture of two barbs. It's not in the *Merck Index* because it does not have mixtures, it has compounds. So you have to go to other reference sources. There is a *Pharmacopeia,* which has a lot of things if the *Pharmacopeia* agrees with it. If they don't, it may be in the *National Formulary*. These are things in the library. I don't hold them as being that important of a thing to have. The *Merck Index* is.

Know where *Chem Abstracts* is. How many people have used *Chem Abstracts*? Good. About four. How many people have heard of *Chem Abstracts*? About eight others. How many people have never heard of *Chem Abstracts*? Good. More than half. Learn the term *Chem Abstracts*. It is probably the first spot to go for searching out a way of getting at a given chemical. Most drugs are chemicals. You'll find that they're listed as a chemical. You'll find a generalized index for people who don't know chemistry and they'll talk about things under common names. But this is just a starting point. It is, after all, a chemical index, and does not involve itself deeply with medical matters, or plants, or mixtures, or phenomena, except to the extent that a specific chemical is involved.

Who's heard of ginseng? How are you going to get at the information? I had a problem that came up. I don't know the answer. I was asked, in fact I was given a book, "You might want to talk about this in your class." It was a book on ginseng. "Hey, that'd be kind of a neat thing. What is ginseng?" "Well, it's one of these things you pull out of the ground in North Korea, if you're cheap, and in China, if you happen to be into the imperial variety. A lot of people swear by it. It's used widely as a drug, and it's one of the major things in the *materia medica* of China. It has a great popular appeal here. It is sold widely; expensively and inexpensively." What is it? Who can tell me what ginseng does? Anybody know? Who's heard of ginseng? Good, I have, too. Who can tell me what it does?

STUDENT: Just kind of makes you healthier, increases your wellbeing.

SASHA: Okay, that's a rough one, but let's hold that in abeyance for a moment. Anyone else want to volunteer ginseng?

STUDENT: An aphrodisiac.

SASHA: Aphrodisiac. Neat.

STUDENT: For asthmatics.

SASHA: Asthmatics.

STUDENT: A bronchodilator.

SASHA: Okay. So. It may make you feel better. It may make you healthier. It may help your sex life and it may help you breathe. I don't know. It's not in the *Merck Index*. It's not in the *Pharmacopeia*. It's not in the USP[1]. It's not in our culture. Where are you going to find out about the culture where it comes from? We have drugs in our culture that that culture doesn't know anything about. How are you going to find out about their drugs in their culture? This book is filled with anecdotes. Here's a story of someone who's walking along and they were saved by

---

1 *United States Pharmacopeia*

having found this root and ate the root and who knows what. You're going to find anecdotes, but what is known in the scientific literature, what we call the scientific literature, about ginseng? To do "research" into the value of ginseng would first require much research into the location of the published literature.

If you read the encyclopedia, you'll find something that was known at the time the encyclopedia was written, and it will probably mention there was a big trade, about the costs, about the counterfeit businesses. I don't know the answer. I've never eaten ginseng. I've seen it for sale. It'll improve your health. Well, how do you establish, how in our culture, how would what we call science establish that ginseng would improve your health? Do you give it to a lot of people who are not healthy? How do you define "not healthy?" Well, I think I'm healthy. Would it help me? I mean, these are basic questions. What is your question? Well, the question is, is there something in ginseng that has medical value? I don't know what's in ginseng. How would you find out what's in ginseng? Probably, you would start by going into the botanical literature looking for what's called pharmacognosy, a very closely allied word to pharmacology, which is the study of plants and how plants have been used in the area of medicine.

But I basically want to get back to the very fundamental question you're asking. There is an argument of strong inference. I've written this on the board: strong inference or inductive inference. How many know the word "inference?" How many know the word "inference" as opposed to "implication?" Fewer hands. Let me try this one. Inference and implication are screwed up more than any two words, probably, in the English language. If there is a message that I wish to get across to you, and I don't want to put it into words, I'm going to make it in a way that you will get that message. It is I who am implying to you that such and such is so. It is you who are inferring from what I say that such and such is so. The source of the information is the implier. The receiver of the information is the inferrer. I imply, you infer, if the message goes that way. You imply, I infer, if the message goes this way. It's a very simple balance of two words, but they are absolutely opposite ends of an arrow of information. One implies to a person who infers. Strong inference is what you infer from what you see.

The concept of strong or inductive inference was brought out by Francis Bacon, who really was the author of Shakespeare's plays. But that's going to really raise more problems than I wish to face. So I'm not going to pursue it. But he was a well-known writer and philosopher of his time, and he was the origin of this concept. I believe it to be a mainstay in what I call the scientific method, that is, devise a hypothesis, something you feel might be so. As I mentioned, the whole area of research is to find fact or principle. A principle is more or less the structure that is built out of the bricks. The bricks are called facts and they go together to form the principles. If you believe the earth is flat, you need a hypothesis. It may or may not be so, and currently we do not believe it is. Devise a hypothesis and an alternative hypothesis. Then devise an experiment that's going to screw it up, devise an experiment that will fail, devise an experiment that will challenge the hypothesis by not succeeding. It sounds backwards. The whole principle is that no matter how often you look at something, how often you challenge something, how many experiments you do, you never can prove that something is so. You can only fail to disprove it's so. A hypothesis can never be verified by an experiment. Its merit can be measured only by the diligence and skill that you can bring to challenge it, for it will take only one inconsistency to bring down the house of cards!

Is this cocaine? Well, I ran a melting point, TLC, GC, and an MS. I ran this spot test, crystal test, right down the line. Everything jived. Forty-two tests said that this was cocaine. Is it cocaine? What if someone ran a forty-third test and it failed? Not cocaine. No matter how many tests succeed, it takes only one test to fail and it's not cocaine. So I ran all the tests I could think of, total diligence, total imagination, vast experience, everything I conceived, it held up with all. I believe it to be cocaine. And probably so. It's an opinion. I believe it to be cocaine. In my opinion, a reasonable person would say, it is cocaine. You have not proven it's cocaine. This is a very important point that demands repetition, no number of successful experiments can ever prove something, but one unsuccessful experiment can totally disprove something.

Let's go and extend this whole strong inference a little bit further. Let's take the argument that a drug shall not be used until it is proven safe. We've heard this, I think, one way or another in connection with popular drug use—such-and-such a drug shall not be approved for social use until it is proven safe. I beseech you to suggest what series of experiments will prove safety. You can prove hazard. A simple demonstration of lethal overdose will do that. You can shove it in a mouse, the mouse falls over, bury the mouse. Hazard! It's lethal! But you don't prove safety. Put it in a mouse, the mouse doesn't fall over, you have not shown safety. The absence of hazard is not the same as safety. You have only shown it's not lethal to the mouse at that level. That mouse, by the way. And that form of the drug, at that time, under those circumstances.

So, design something that will tear your hypothesis apart and do the experiment. If it doesn't, cycle around again. If it does, cycle around again. This is a process of inductive or strong inference. One of the things I like to do is I can usually find a good, aggressive person in my criminalistics class, give them a quarter and ask them, "What is it?" At the beginning of the semester, they'll say, "It's a quarter." I'll say, "I hope by the end of this hour and definitely by the end of the semester I can get you to say, it appears to be a quarter, or I think it is a quarter." The issue here is that this person will eventually be appearing in court as an expert witness: "I was given what I took to be a quarter, what I believe to be a quarter, what appeared to be a quarter, what I assume to be a quarter because what it had was all the properties of a quarter." What are the properties of a twenty-five-cent piece? The idea is, there is an exemplar. Exemplar is the thing against which you compare everything. It is the thing which is the standard quarter that everyone compares the quarter to. It's the thing back in the mint in Philadelphia. It's like the platinum meter that sits over in a museum somewhere in Paris. It is the meter against which all other meters are compared. That is the standard meter. It is the exemplar.

The student who was doing the work with the explosive residues went happily on, compiled the literature, and turned in his master's thesis, which was largely a literature review. He shot one gun, gathered a few things that showed the targets were there, and is off doing something else now. The idea of asking a question, designing a question, making a question sufficiently narrow that's within your capacity to answer, is an aspect of research. But knowing about what is already known is very much a large part of research.

A very important point in how you pursue this through the library. How many people are familiar with *Citation Index*? One. That's not enough. We need a diversion on how to use the literature. Anybody who's going into the scientific arena at all must know this set of books. Do you know if they're in the library? We'll see. You will never be able to find out everything that has been published. So how do you find out good starting information on what's been published? Let's take the example I gave of ginseng root. How are you going to find out? Well, if you can get to one paper, one book, one citation, anything that mentions what you have in mind, you have a place to start. Let's say you find a study that was published in China, but in English, in 1968, that gives a clinical study on the influence of ginseng root as an aphrodisiac to see if it increases libido. So here you have a study. It's a starting point. You have an author, title, citation, where it's located, the year, the volume, the page. So you have a starting spot. Everyone, I think, intuitively knows that you can go to that paper and go backwards in time. You can go to the bibliography of that paper and look at the references that talk about the same subject and all of these predate your current source, this is going backwards in time. Then you can go to each of these citations, if you want to, and find out whom they cite. And each of those, in turn, have cited people. So here is your paper, in essence, you're forming a tree [drawing on board] of the citations from each paper. And pretty soon, you'll find they begin to overlap, and you find a body of literature on the subject that is the basis on which you can ask your question sensibly. You know what's known. This is the bibliographic foundation, the citation literature of what's there. This will bring you up to the time of that paper. This is straight bibliographic searching, easy to do in a library.

The *Citation Index* allows you to go the other way and come from that paper up to the present time. In the same concept but intaglio, or sort of inside out. What published papers have cited this paper? Well, it may turn out there are fifteen papers, from then to the present time, that have actually drawn citations to this paper. What papers have cited those papers? Well, there are five papers that cited this, four that cited that, two that cited another and seven that cited another one. So you are going into the past by what papers it cited and into the future by what papers cited it. It's a beautiful way of really addressing a situation, of writing a review article on a particular area and you happen to not quite know how to get there. Get another paper that's not in this network but is somewhat related, get a third paper that is somewhat related but not part of the network, and follow them back in the past, and then follow them up to the present. And you can really get a picture of where things are.

This is certainly a way to follow the example I gave earlier concerning Ginseng root. Let me go to the *Citation Index* and find all papers in, say, 1982, that cite the specific book I had, with the anecdotes narrated in it. Amongst those papers I would expect to find at least one scientific review or research paper that dealt with the chemical composition or medical use. Then I can go backward in time from that paper by using its bibliography to find things it has cited, and go forward from that paper in time by using the *Citation Index* to find papers that cited it.

And use a second and a third paper as starting points—papers that may not be in the above upwards and downwards trees because they are really only somewhat related. And pursue them both ways. Pretty soon more and more of the citations will begin to overlap, and you will find that you are building an ever-tighter network of what is known about your immediate area of interest.

You'll find, if you're at all diligent and interested in that subject, that it's two o'clock in the morning in the library and they've locked you in. It really gets exciting, and suddenly you realize "I didn't know that!" and you are in the process of learning what is known! I've been in research for forty years. I love it! I'm deeply invested in it. I love getting new ideas. I'm continually amazed by it. I have a whole file at home which I call "Oh Wow!" where I put things as I find them that might be totally unrelated. I found what the active principle was in the cashew nut that makes the hand go itchy. It's right next door to what's in poison ivy, oh wow, except it has a phenyl over here, oh, wow! I don't know if they're going to make any sense out of my research. I don't care. It's all a part of getting information, getting it, not as an end in itself, but as a contributor to your asking questions.

That is a major, important portion of all research work, finding out what is known. Finding out where to find out what is known. I mentioned the *Merck Index,* and the *Chem Abstracts.* The *Biological Abstracts* is almost worthless; it is not thorough, and is not particularly up to date. *Current Contents* is excellent because it has keywords in the back. How many people are familiar with *Current Contents?* Not many at all. This is something that you should become familiar with. Published once a week by the same institute that puts out the *Citation Index. Current Contents* publishes the title pages and table of contents of every journal in that discipline every week.

ANN: Is that in the world or just this country?

SASHA: In the world. The titles are largely translated into English. But the things you may want to find are in there every week, often ahead of the appearance of the actual journal in the library. Journals often send the table of contents to this ahead of time because they can pull it out of the rushes, out of the galleys, and send it out so it can appear ahead of time. Then, once you have the journal you want, you look to the author and in the back is the name and mailing address of every author who's in every journal that's been published every week. So you can write to the author. Get the phone number and call them if it is something you are doing and you want to get right at it. "How'd you find out about that? I just signed the paper and it went into the galleys." "I got it from *Current*

Contents." "Oh, good." Then, you have a very important thing called the keyword index, which is every word that's in the title that is not "an, of, to, by" or what have you, is put in the keyword index, including monstrous chemical names. The names of stars, physics, chemistry, all these disciplines, there are five or six of these, are in there too.

Become familiar with it. I spend an hour a week in it. I could not succeed in going to a library an hour a week. All I do is read through the table of contents, I snag a book or two, I take a half a dozen journals home, that's all. Enough in the very tight discipline in which I'm actively publishing. You get to a point where you're kind of in an old boys network. You know who's publishing, where they're publishing. They'll tell you when they're publishing. They'll inform you; it is kind of like the mycelium of a fungus that spreads through the community that you can touch and know what is going on, and people let you know what they are doing. An hour a week is very well warranted in this direction if you're working in a general area in chemical research. Find out what's been published. Find out what's going on. A lot of it is hopeless trivia, what I call observational reports. Like pharmacology which is an art and a science at the same time.

One person once described pharmacology in a very nasty way, but there's a touch of truth there; pharmacology is the art, or the science, in which you inject a compound into an animal, and it produces seven publications. [Laughter.] Okay, a little bit nasty, but a little smell of truth in it because a pharmacologist will not often ask, "What is the property of the compound?" They'll ask, "If I put this compound into an animal, what will it do to the animal?" And if it makes the animal up or down or stars or sideways or dead or nothing, the pharmacologist has gotten a response. The compound, surprisingly, a close ally of a very well-known neurotoxin, is totally inactive in the rabbit. [There's a] paper. I mean, you don't even have to have positive results.

I got into this area in a very strange way. I've been trying to resist drawing dirty pictures on the board, but I will. During my very first job in research in industry, I came into a company that was able to make in quantity a very unusual compound: meta-t-butylphenol, which is a white solid that had never been known before. They had ways of making it by the bucket. And their question to me was: what could it be used for? I didn't know. I knew that what popped into my mind was the fact there was a well-known compound that has a structure that's known as physostigmine or eserine. We talked about the sympathetic nervous system and parasympathetic nervous system. This is something that activates the parasympathetic nervous system. It's a rather good insecticide. It causes a transmission of signals across the parasympathetic branch. It acts as if it were a choline compound.

I knew about this from playing around with alkaloids some time ago. I never worked with the compound, but I knew about its structure and I knew basically that it was a key to a lot of insecticides. [In reference to what was drawn.] What I saw in it was, lo and behold, if you look at what's inside that circle you have that compound here. I'll bet if I were to take this compound and since that one has a nitrogen on there, I better put a nitrogen on here too, and if I made a carbamate of that phenol group, you'd have an insecticide you can make by the ton. And industries love things they can sell by the ton. I did one magic thing that really was a life saver. I wrote down my ideas and had two innocent people witness them. I said, "I bet if you were to put a nitrogen on there and a carbamate up there, you'd have an insecticide, John Smith, Joe Blow." Sure enough, it made an insecticide.

Not only that, it turned out a very close ally of it went commercial. And it's one of the reasons that I was able to work doing my studies at an industrial company for a few years. They said, "Well, if you can predict from theory something that would go commercial, you just go right ahead with your theory." And I said, "Yowzah, boy! I'm going to go right ahead with it." And I worked for ten years before they finally realized the area I was really interested in, which is drugs that affect the human mind, was further and further from what they wanted to go into. They loved insecticides and fungicides and things that killed crops and didn't kill crops and so forth, which is not my interest at all. But that kind of projection came from knowing that these kind of structures existed and what those structures were, and looking at that thing and being able to turn things around in my mind and say, "Oh, they're very similar." The name of the compound that went commercial is Zectran. It sounds like a motor additive, but it bought me a few years of fun research.

That is the question, what can you do with it? Well, that kind of question you can answer. They didn't say, make a wheat preservative out of it. Which is the sort of thing you may not ever get to do with it. The basis of questions in research is how you go about converting ideas into products. This is the other half of research that everyone thinks is all of it, and I want to talk largely about that, which I have about half the hour left to do. I did give a handout with a lot of boxy things on it, which is more or less the process by which drugs that come into usage get their origins. Although, you can buy a drug by going to a drugstore, how did the drug get to the drugstore? It came from the manufacturing house and they devised this through their research policies. Where do drugs that come into usage get their origin? Where do they actually start? This is the question I'm addressing in this handout. Almost every drug that has gotten into human usage has, at one time or another, started fundamentally in human beings. It has either started in human beings or it started from a screening process that has been based upon a drug that has been started in human beings.

Let's take, for example, the company I worked for, for ten years, which was Dow Chemical. It was kind of an interesting chapter on a number of levels. They were into industrial chemicals, different kinds of largely agricultural work. But my interests were primarily in the area of drugs that affected the human character. One of the drugs that I actually devised that did go into commercial study for a while was a drug called dimoxamine (Ariadne). It was an antidepressant. How do you find an antidepressant has a biological activity? Well, you take a person who's depressed and find they're no longer depressed. You can't do that in a very easy way, so you set up animal screenings. You have a host of animals. Now a lot of animal studies are easily run. If you have a fungicide, you take a colony of fungus and see if it is easily killed. If it's an insecticide, you put a colony of insects in there; does it kill the insects? When you get into human beings and animals, it gets a little bit trickier because you have to have something that will kill the fungus on the host but not get the host at the same time. A marvelous way of killing a fungus is with a blowtorch. It is one of the best treatments for killing fungus, but it is hard if you have athlete's foot because you get burns out of it. It sounds silly, but you have a basic balance between getting at the biological activity of the guest without harming the host.

On this handout, the animal pharmacology really is the source of the drug. In industry, a lot of the effort is conducted by people who are taking a lead (the leads of course have come from drugs that are known to be drugs) and synthesizing things that are close to it, variations of it that spin two things together, that exclude certain things, making compound after compound, sending them down a pipeline, at the other end of which there is what's called a pharmacological department (maybe next door, maybe across the country). There, these compounds come in, they are looked at, and they're put into animal screening.

For example, here is a set of animals and we put electrodes to their heads every day. We can handle twenty of them a day. We push the switch on the electrodes and they go "blah!" They electroconvulse. And then we'll put a compound in each of these twenty animals, we'll put the electrodes to their heads, push the switch, and see if they don't convulse. We might find an antiepileptic, something that keeps animals from convulsing. So you get an animal model, a model of a convulsion, a model of an animal that is doing this or not doing something else. And you will test these compounds in each of the animals, find the leads, and down the pipeline comes the answer: this is the most promising of your ten compounds. Okay, we'll make ten that look like it. Down the pipeline it goes.

You very rarely have in the same laboratory a person who is the source of the material and the evaluator of it. You have this separation. The person in the laboratory loves having to work with this type of chemistry. The person in pharmacology loves this type of action. They're not in agreement. There's a conflict. There is a committee concept to much of research that is a great limitation to what is actually going to be discovered. What is discovered that becomes new is often by the person in the laboratory doing pharmacology or the person in the pharmacological lab doing chemistry. They are pursuing something for which there is immediate feedback.

Take, for example, how you define new sweetening agents, agents that you put in coffee that make coffee taste sweet. How would you go about finding them? It's your job. You're hired and you are working for Monsanto. "Find a new sweetening agent. We want to knock Nutrasweet off the market." How are you going to find it? You're right now at the nitty gritty of research; your task is to find a new sweetening agent. Here are our leads. Here are five materials that do cause sweet tastes, but this is too toxic, this has a bitter aftertaste, this one takes fifteen minutes to come on, this one causes cancer, and that one causes teratogenesis. We can't use them. But we need one because we're losing the market. Saccharine is not going to be available much longer. How do you find one?

Well, my philosophy, that people would cringe at, is to put a damp finger into it and taste it. [Laughter.] That to me is the heart of how you find a sweetening agent. Well, what if it's going to cause cancer of the jaw? Okay, then you come down with cancer of the jaw, but you've found a sweetening agent. [Laughter.] So you have risk and you have reward.

If you begin training an animal how to respond to a sweetening agent, you're training the animal to respond to your viewing of how the animal responds to a sweetening agent. We're going to put it in the water and see if the animal likes the water. The animal may like bitter. You don't know. So you may come up with the world's best bitter agent because the animal is responding and relates sugar to bitterness. You don't know what they think so you use rewards to learn that. The animal will respond if you give sugar in this water and not in that water. When you put an unknown in a sample of water, will they take that in preference to sugar or not? These kind of multitask experiments are very commonly used. It's a hard, hard task.

Let's say for some reason you find that the oxime of some terpene that comes out of a pine tree does taste sweet. How are you going to begin getting this thing into the market as a potential sweetening agent? Well, that's more or less what this handout describes. You have the screening in animals. And then you have foreign sources. You have people with patents overseas. A major lifeblood of the industry is getting a claim on something so that no one else can do it. There are two types of patents for protecting inventions. One involves the ownership of the compound itself (composition of matter patent) and the other is for the newly dis-covered use of a known compound (utility patent). The big powerful one is the composition: "This material is new, and I am claiming it to be new. It has a couple of uses. That's incidental. It's new and it has uses which are part of the new thing, but it's a compositional matter. This oxime of the ketone from the pine tree has never been made before. It happens to be sweet, but I'm patenting it for anything else." Maybe five years later, it turns out it's the world's best fuel additive. Whoever discovers the fuel additive can't patent the compound because it's new and you have the patent. What you want, if you can, is to get a composition patent. If it's a known compound, no go, you can't do it. It's already known.

Say you've found a new use for it. In order to limit the number of copycat compounds potentially made in the future by a competitor, you make the use patent as broad as you can. The original claimant will try to stake out as much chemical territory as possible. Some applications will contain broad groupings "with a chain of from one to eight carbons in length" and "with X being any element from the periodic table." I once saw a patent issued to DuPont, which, if you took all the permutations of R1, R2, R3 on an indolic structure, I figured out came to roughly 10^9 or 10^{10} compounds that were implicit in that patent, which is 100,000 times more compounds than have ever been made in the history of science. But if they were in that patent, they were technically patented. But that patent would be destroyed in a court case.

[Directed to student] Yes.

STUDENT: I heard a lot of generic drugs come in because patents expire.

SASHA: Expire. Patents are worth eighteen years.

STUDENT: So patents have a certain time.

SASHA: They have a certain time. I think I may have mentioned Darvon napsylate. Darvon is a good example. It was patented about twenty years ago and its patent expired eighteen years later. You have eighteen years, you cannot renew, it becomes a public thing. But often what a company will do is say, "We've got the name trademarked. You can't use the name. But what we'll do is find a gimmick that allows some new form, some new novel, inventive change." In the case of Darvon it was Darvon napsylate or Darvon N. The "N" meant that it was now a new salt and the salt had a property that Darvon itself and all its other salts did not have. Namely it is not very water soluble. In fact, it is very insoluble. And the result is you could take a big bucket of it and you'd only get a little bit of effect over a period of time. Darvon N was a time-release spansule kind of thing, so it was effective for a longer period of time. You took more of it and it was slowly absorbed into the body. So they patented Darvon N, put a big advertising campaign out, and had another eighteen years. So, you can modify it, but the life of a patent is eighteen years.

So, patents are the lifeblood of a company. When a new drug comes out, you have a copycat approach through all of the industry. It is commonly thought that there are vast armies of researchers continually making new compounds, and they are injecting them into battalions of test animals, looking for some response that will suggest a potential commercial drug. In truth, most of the synthetic effort is directed to modifying known drugs for purposes of patenting or of improving effectiveness and safety. You find this in the barbiturates when they first came out. You must have 200 to 300 barbiturates that are patented, only about half a dozen of them have different properties. Many of them are merely so they can get in on the business. The benzodiazepines, the whole Librium/Valium area, extraordinarily broadly patented by many different companies. But all basically from imitations of a given thing. Take the steroid area. So-and-so patents a steroid. Word comes across that it's effective in clinical work. "We don't have an anabolic steroid. We've got to have an anabolic steroid." Into the research department, crash program to get an anabolic steroid. "This is known. Mike, put a fluoro group in the four position. Jill, put a fluoro group in the five position. Sally, you take care of the sixth position. Henry, you take the seventh position." Everyone begins synthesizing all these and they go into screening. Pretty soon, two new compounds come in and go into pharmacological study. This is getting down to the animal toxicology and metabolism. Still, in animal work. Find out if it's going to be poisonous or if it has long-term effects.

The necessity of animal study is extremely important, especially now that so many compounds have been found to have long-term subtle changes in the health of the person who uses them. Put them in animals over three generations and see if the offspring of the offspring of the offspring come out with the right number of legs or proper health or have the proper lifespan. Find out what the compound does. Make it radioactive. Inject it into a test animal. See what the metabolites are and find out if the metabolites are toxic or have bad effects over generations of animals. The drug may have metabolites that are themselves possible drugs, and may provide leads for further research. The antidepressant

Desipramine is an active metabolite of imipramine, just as nortriptyline is to Amitriptyline. And the drug may be a mixture of things. These may be impurities, and each must be evaluated as a separate entity for its toxicity and consistency. Make the compound. Find out it's only 92 percent pure. What's the other 8 percent? Can we get rid of it? Characterize it, find out if it has properties over three generations of animals.

Some of the materials that are on the commercial market are hopelessly impure. One of them is—I can't remember the name—you add chlorine to a terpene until it's 68 percent chlorine. It's sold as an insecticide. I think it's toxaphene. It's a chlorinated terpene. Terpenes are a mixture of hydrocarbon compounds. When you take these and put chlorine in until it's 68 percent chlorine, you have some 177 chlorinated hydrocarbon compounds that have been characterized. The mixture has been sold as an insecticide. What in the world is its pharmacology? I would imagine half those compounds are not even known. They're just bumps in a chromatogram somewhere. So this is the kind of thing that cannot occur now. To be patented it has to be substantially pure, totally pure, 100 percent. Strong inference of the identity will not do it.

But with those things that are impure we have to know what their properties are. Make them. You have to know what the source of making a compound is, and then you have to have the recipe for getting there pinned down and put on ice. Because if you make it by a different way, you're going to come up with different impurities or different degrees of impurities or different strengths. All these things must be done before it ever even goes into clinical trial. So, you have invested at this point, this is the break point, before we go into phase one of clinical drug trials (right across the middle of the page in the handout), you have now probably invested ten million or fifteen million dollars into the compound before it goes into human beings.

This happened with the compound dimoxamine, which I mentioned earlier, that I got introduced into Bristol.

[Directed to student] Yes.

STUDENT: Was that a tricyclic, similar to—

SASHA: No. It's very much like the amphetamines. It's an amphetamine without the stimulation property. Quite a different class of compound. It's very close to a hallucinogenic, but it did not have any visual effects. Hence it had the virtue of being potentially clinically useful. Any company that gets to this ten million or fifteen million dollars place of now being able to put the compound into human beings has to go to the FDA. Take this ten feet of paperwork and say, here are our animal studies, metabolism, what we are going to name it, patent coverage, toxicity, animal demonstration of efficacy, our demonstration of a lack of hazard, who we are going to have distribute it, we are going to call it such-and-such, and request to test the drug in human beings. So you get what is called an IND, an Investigational New Drug. This is permission for it then to go into the first of four phases of clinical study. Phase one, is it harmful to humans? Let's say you have a drug that is potentially an antidepressant. It is nice to know if it will treat depression in humans, but first you want to know if it will take a normal person and do something wrong to them.

This is phase one, where you're taking a drug into a normal population with all this background work, millions of dollars invested, and finding out if there is something intrinsically wrong with the compound: interferes with sleep, interferes with appetite, interferes with any of the normal patterns of that person's behavior. You have a clinical study, maybe half a dozen clinical studies, in which you will then amass ten or fifteen people per group, maybe 100 to 200 people, where they take blood pressure, long term this, long term that, complete body chemistry, the works. Then this goes back to the FDA to say "We've gotten through phase one. Now we'd like to apply for phase two.

Phase two is going out into the abnormal population and performing tight clinical studies, the population that is lacking what this drug will supply, or is deficient in what this drug might repair. If they are people who are depressed, you'll get it into populations that are depressed and see if this is an antidepressant. Again, a very limited study. Phase Two

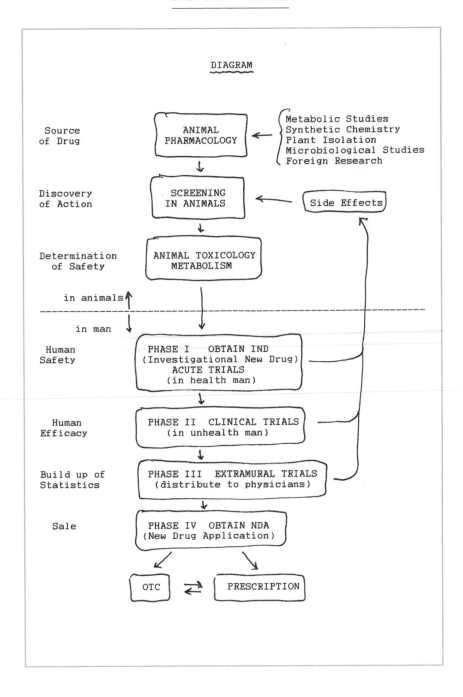

DIAGRAM

Source
of Drug — ANIMAL PHARMACOLOGY ← Metabolic Studies / Synthetic Chemistry / Plant Isolation / Microbiological Studies / Foreign Research

Discovery
of Action — SCREENING IN ANIMALS ← Side Effects

Determination
of Safety — ANIMAL TOXICOLOGY METABOLISM

in animals

in man

Human
Safety — PHASE I   OBTAIN IND (Investigational New Drug) ACUTE TRIALS (in health man)

Human
Efficacy — PHASE II   CLINICAL TRIALS (in unhealth man)

Build up of
Statistics — PHASE III   EXTRAMURAL TRIALS (distribute to physicians)

Sale — PHASE IV   OBTAIN NDA (New Drug Application)

OTC ⇄ PRESCRIPTION

clinical trials to unhealthy humans. Almost always this is contract work. Here is a group that makes its livelihood by being in association with a clinical study of chronic depressives in such-and-such hospital in New Jersey. They are in the business of doing phase two FDA studies on anti-depressants. You give them the compound, you give them 100,000 bucks and they'll give you the results on fifteen patients over such-and-such a period of time. That's their livelihood, that's the industry's livelihood. It is not a matter of we bought and sold this, it is a way some people do business by making compounds and some by evaluating them.

Phase three. Once everyone is satisfied that there are no long term negative effects in the reports that have been coming in, phase three is where you send it out, outside of your control. Give away for free, take this, a new experimental anti-what-have-you, try it out in your patient population. We will provide the drug to you and all the information. We want back from you all reports on it being used. So its use from the field, where there is not a regulation on how it's used, but there is a demand. It's not yet open for sale, not yet open for availability. It's this much more widely broadcast experimentation.

Then phase four is actually the development of a prescription or an over-the-counter status.

STUDENT: What's phase three?

SASHA: Phase three is these clinical trials by physicians out yonder, outside of the control of the company.

STUDENT: You have a phase two clinical, managed by the drug company.

SASHA: Yeah, phase two is under the company's control. They contract a little clinical study.

STUDENT: Right, and then phase three—

SASHA: Phase three they distribute to physicians, whoever wants it.

STUDENT: Free samples.

SASHA: Free samples. Mmmhmm. After phase one, two, and three, the compound can be approved and marketed either over the counter, OTC, or with a prescription, having a physician sign to allow you to purchase a drug. Phase four would be after that, to follow up on potential unexpected adverse reactions, but not all drugs require phase four.

STUDENT: So phase three, that's where you get your sales in, go around to different doctors and pharmacies saying you've got this new—

SASHA: But the requirement is not quite license free, because they have to send back reports. The forms come along with the drug. The detail man is usually pushing a new drug that has been approved for marketing after Phase Three, and is trying to get physicians to try the new approved drug.

Now, how much damage, how much hazard can you afford, how much risk, how many toxic things can you afford to compensate for the virtue that the drug will provide? This is one of the basic questions in pharmacological research. It's a hard one to address. An antidepressant called Monase was introduced to the market about ten years ago by Upjohn. It was an antidepressant, moderately effective. One person in about 5,000 had jaw problems. The jaw would tend to lock, and would tend to have problems in the jaw motion. And it was considered not acceptable.

Take a compound such as chloramphenicol. An extremely potent antibiotic. Very simple compound. Very potent antibiotic and, against certain diseases, effective where very few drugs are effective. It causes blood dyscrasia in a certain small percentage of people. But in that case, the difficulties with the blood problems are outweighed by the fact that it will get at certain diseases that cannot respond to other things. So you weigh a really intense virtue against a risk that is, in the balance of things, not sufficient to discourage its use. The use of that kind of antibiotic indiscriminately is irresponsible because it does carry risks.

If it's used, for example, for treating colds, that's a poor use of it. Almost any antibiotic is of questionable value in that case because colds don't respond to antibiotics. Colds are not caused by bugs.

All drugs have risks. All drugs have side effects. Who's heard of the PDR? Who has not heard of the PDR? Okay, this is going to be a course in three letter acronyms. [Laughter.] *Physician's Desk Reference*. A book about three inches thick. Comes out every year. It's given free to every physician as an advertising procedure. I think its commercial price is twenty or thirty dollars. It lists every prescription drug that every company puts out. It has a bunch of fold-out colored slides that show you the picture of the drug and its code. So if you get a prescription pill, you can trip over there to where it is and find out what the material is, the color code of dosage, the things for which it's used. And it has a blurb under the drug that tells you what it does, how it's available, the size dosage it has, and if it has warnings. And all this information is published by the drug company itself. It's a compilation of the inserts that come along with prescription drugs, in a single place.

Physicians don't particularly like patients to have the PDR because it tells all the side reactions. Not the side reactions that may be commonly seen, but any side reactions that have ever been reported. You can go on for three inches down the column of side-reactions. Just like having *Your Health* or *Medical News Today* in the waiting room of a physician, the patient won't find it there. Instead, they have *National Geographic*. Because you have a person waiting who doesn't feel well. If they open up *Medical News Today*, guess what their symptoms are? The first ten things they read and they realize they're coming down with the extremely rare whoopy rami jammies that no one's ever had since 1912. But they come in presenting with a marvelous set of syndromes. The poor physician says, "I thought you had a sore throat?" The mind is a very active thing, you don't want to necessarily feed it all these things.

Talking about the mind being very reactive, I just had an interesting incident occur in a meeting yesterday where we were figuring out what went on in our current tobacco research in the differences between people responding to our intravenous injection of nicotine. It has people who are smokers and some who are in abstinence and who are getting intravenous nicotine to find the kinetics of a single bolus. You introduce the nicotine, you find out what all the dynamics are. Will it interfere with the urge to smoke after a meal? One of our current problems right now is, apparently (I was not aware of this kind of a statistic or information that had been gotten epidemiologically), those people who smoke often have an urge to do so after meals—how many people actively smoke? Okay, about six. I would ask each of you in turn, if you were to be cut down to one cigarette, you're allowed one cigarette a day, that's it, which would be your favorite cigarette? When?

STUDENT: Last one in the pack.

SASHA: Last one in the pack. [Laughter.] The favorite.

STUDENT: Last one before I go to bed.

SASHA: Last one before you go to bed. Who else smokes? The favorite.

STUDENT: Morning.

SASHA: Morning. Another.

STUDENT: After a meal.

SASHA: After a meal. Another. Anybody else? Smoking. Any other ideas? My answer was the first one in the morning. Most effective. Wack! Whoooh! After that it's all downhill. After a meal is apparently the most favorite place for that one cigarette. Why? What is it about a meal? And so, a little research question. The question asked was: "What is it about a meal that makes a person want a cigarette?" Hypothesis: Maybe the amount you wish to smoke is dictated by your nicotine level in the blood, and as the nicotine level drops down you want to balance it up with another cigarette.

So you may be dictating your smoking habits by nicotine level. You wake up in the morning and it's way down there. Bounce it right up and away you go. After all, you get food in this way. The hepatic portal shifts the bulk of the blood flow down to the gut which takes it away from the brain. That's why you get nice and sloppy and sleepy and happy to sit back and relax after a meal. Could it be that you're processing more of the nicotine through the liver that tends to chew up the nicotine and hence drop the nicotine level faster after a meal? That's the hypothesis. I really got off my point, but I'm getting back to it. [Laughter.] Here's how we set up the experiment. We stuck a needle into their arm with a little motor that pumps nicotine into them.

First of all, let them have two quick cigarettes. They've been abstaining for about three days. Then, put a needle in the person's arm and maintain the nicotine level in them until it is stable. It takes about three to four hours to get a feel for stability. The person surprisingly does not particularly want to smoke. I think possibly because they have tubes going into every part of their body. [Laughter.] But anyway, the blood level of nicotine does the following: it gets fairly static. So you have this kind of a thing over a period of time. Then, after about four hours, have them eat an 800-calorie meal.

Now, one of the two really ugly parts of the experiment was eating an 800-calorie meal in ten minutes. That was really objected to. It was not comfortable. People really rebelled at that.

[Directed to student] Yes.

STUDENT: Couldn't they give them a semi-solid shake?

SASHA: Could have. But you'd not know what you were looking for as we were looking for the most global imitation of real life. You can only try to get as close as you could with tubes and needle, and as close as you could with the meal. The hot thought was, "Maybe the nicotine level will drop, despite the fact that you're infusing it." What the nicotine level did was to not drop. There was no change in the nicotine level right after the meal. Beautiful hypothesis. Do it, and throw it out! You do not have a nicotine level drop during the meal. Is the urge to smoke due to nicotine levels? We still don't know. Is the reason for smoking after a meal something entirely different? Probably. It's not due to what we had hypothesized.

[Directed to student] Yes.

STUDENT: Wait. When this test was being done, they were constantly getting nicotine?

SASHA: Yes, continual infusion.

STUDENT: But when you eat, you don't smoke.

SASHA: That's right, you're not smoking for this whole four-hour period because all the nicotine is being supplied by a tube in your arm. And it's being pumped in.

[Directed to student] Yes.

STUDENT: Sorry. If this were a real live situation, though, the person would be deprived of nicotine during that time when they're eating. And so, it would automatically drop because they were eating for that ten minutes—

SASHA: During the time of eating. And maybe during the eating period you are actually giving a longer period of time than you would normally have gone without smoking. I mean, you smoke every now and then, and then you stop smoking while you eat.

STUDENT: I know I've been at long meals, about seven course meals, without a cigarette in between—

SASHA: So it may have nothing to do with eating. It may be that you're just getting away from the ability to smoke gracefully at a meal. Good argument. Still, more people than not say that the cigarette they prefer is after the meal. Your answer may be absolutely correct because that is a longish period that has been deprived of cigarettes. I don't know the answer. But this was a specific one that could be addressed: is there an actual change in circulation? And the answer was no.

[Directed to student] Yes.

STUDENT: Why did they use 800 calories?

SASHA: For one thing, you want to get into a reasonably short period of time with the eating mechanism in case there is a saliva, hormonal, or other response that comes with the process of eating. You could have given a bolus of the calories in an injection; say, give them a pile of amino acids and sugars. But maybe there's something in the mechanics of the eating that matters—I don't know, except that I know I don't know the answer. How would you design the experiment? Would you have had a longer period of time or more calories?

STUDENT: A longer period of time.

SASHA: Which would make change less visible. You want to scrunch as much as you can so any change will be most amplified. If you spread the time of eating over the course of four hours, you would not see as much of a change, even if it were there. So you want the period as short as possible to get the most amplification on the effect. And as many calories as you can get down in a short period of time. Again, that would imitate normal eating patterns.

There's a good question that's now being addressed, "Do people put on weight when they stop smoking?" The answer is "Yes." On the average about ten pounds. Some people not at all, some people twenty pounds. Why? No one knows the answer to that. There's a good question. How would you set up an experiment to determine why people put on weight when they stop smoking? Well, you've got to look at the whole experimental situation. You've got to get people who smoke and are willing to stop. You've got a small population right there who are going to be willing to stop for a small period of time. In our studies where we have people stop smoking for four or five days, an amazing number of people cheat. It's an amazingly incredible addiction.

We're going to have a whole section just on tobacco where we'll talk about what's going on in the body. But the question is not known. People say you tend to eat more. Well, that's ridiculous. Do you tend to eat more? If you put on weight, you may very well be eating more, but do you eat more? Well, no one's actually weighed the amount you eat or weighed you after or before meals. Perfectly good hypothesis. Maybe you're screwing up your hormones and the hormones that normally regulate appetite are getting all screwed up. The answer is not known. Maybe if you could actually find out, then you'd have a point that might be addressed to help people overcome that particular addiction. But this is the whole problem of asking the right question in the area of research.

Often the side reactions that are found during the course of this Phase One, Two, Three, Four of the FDA approval are the clues toward new drugs. A good example of this was the study of drugs that were being used as antihistamines. This was done by Jean Delay in France. I think it was in the early 1950s. They were studying a whole series of antihistamines and amongst them was a material known as promazine, which is a big three-ring rascal, so its development was the outgrowth of an antihistamine study. They were studying a whole bunch of asthmatics in this ward, and they were trying different analogs of promazine, which in animals showed an antihistamine property.

Delay was a good observer, and he noticed the ward on this particular study was dead quiet. There just was no noise. Usually there was chatter, chatter, chatter. It turns out that this material wasn't a very good antihistamine, but it quieted everyone. That was a side-effect that had nothing to do with antihistamine. A little chemical exploration by putting a chlorine group on and it's even more effective, now it's chlorpromazine, which is Thorazine, and was the first of the antipsychotics, the quieting major tranquilizers. It was discovered because it had a side effect for people who were looking for antihistamine action. And that opened up a whole new chapter, but it was discovered in human beings.

Take, for example, a specific compound I mentioned, the dimoxamine that went into Bristol Laboratories as far as phase two as an antidepressant. A compound that I worked out some years ago. I gave a talk and discussed its action in the human study that I had run in myself and in others. And I said I believe it to be an antidepressant. It was explored first in human beings. It went into the industry, at which point they spent the whole first half year finding an animal test that would give antidepressant action. How do you determine anti-depression in an animal? Well, it's a dicey thing to do. What you do is take a bunch of compounds that are known to be antidepressants and get animals that respond in a certain way to these compounds and then slip this one in and see if it does the same sort of thing. It doesn't address the question, but it does give you that approach. Using known compounds for setting up your animal screening.

Anyway, they did find this test. But before they launched into a full multi-clinic study to determine if it's going to be worth the animal studies or not, every person on the board of directors took it. Which means it was tried in human beings first. The FDA doesn't know about it. There was never any report on it. But the human evaluation of something that is psychotropic, that changes the state of the mind, is substantially the only way you're going to document if it has that kind of action. I don't care how many animal tests you have; the human is the experimental animal that discovers most drugs.

Lecture 3   (SFSU, Spring 1987, Chem. 106) (meant to be Lecture 2)

THE ORIGIN OF DRUGS

     To go to the origin of drugs, it is necessary to go back to
the origin of man.

     I have always had difficulty visualizing myself in the role
of an observer, maybe 5000 years ago, when there were pyramids
being built in what is now known as Egypt, and hunters lived in
caves in what is now Europe, and nomads lived on horses in what
is now Mongolia. The problem probably lies in that I am trying
to place myself in that historic setting with the burden of
having the knowledge, the intelligence, the sophistication, that
I have now, today. The concept of going there, returning here,
as if in some sort of a time-machine, ignores the fact that the
transition from then to now has been an irreversible change.
Every time anything occurs, it changes the human consciousness,
the human perception of reality, and that change is there
forever. There is no way of undoing it and going back to "see
how it was before."

     The entire process of human history has been one of
continual, irreversible change. This is just as evident today --
something happens that affects us negatively on a personal level
or on a global scale, and much as we would like to say, "Can we
go back and try it some other way," and then un-say harsh words,
or make a president undo an act of war, it can't be done. We
cannot re-write what has become part of the record of history.
Every event, once it has occurred, is on the books forever.

     But try something even more difficult. Go back in your
imagination, not 5000 years, but back to the origins of man,
because that is where the story of drugs really started. Back
500,000 years, a million years? I don't know, and no one knows,
as there is no record. One can only guess what might have gone
on then by observing the life patterns of aborigines, or
deciphering the myths and legends maintained by oral traditions.
There are groups that can be studied, or that have been
studied within our short period of record-keeping, who might be
acting out their lives with only minor changes from the earliest
times.

     Let's make it, for simplicity, a million years ago. There
was the earliest man, probably somewhat shorter, possibly with a
lot more body hair, and with only the most rudimentary ability to
talk. Something had occurred. He had become something quite
separate from the ape, or whatever not-man preceded him.
Whether this occurred by evolution, or by some "divine"
intervention makes no difference whatsoever. Something
occurred. It may have taken place in a moment, or it might have
taken many generations. It may have been at one location; it
could have been in a hundred locations.

     What occurred was the ability to ask "Who am I?"

1

With this came the recognition of ego -- that there is an
"I-ness" in me. And with this also came the awareness of death.
And the need to communicate about things other than the
instinctive activities such as food getting, mating, and safety.
Try to put yourself, at your own age, into this environment.
There is no fire, except that generated by lightning, there are
no clothes in the warm climates (in cold places, you use the fur
skins of dead animals), you live in caves, and there is a very
hostile world around you. You must continually seek food, and
you know that you, yourself, are regarded as food for certain
animals. You cannot be damaged, because if you are, you know you
will die, and you are afraid of death. You ask "Will I die?" and
you know the answer is yes. And you ask, "Why will I die, and
where will I go when I die?" This curiosity is unique to man,
and this time, perhaps a million years ago, was the first time it
had ever occurred. There was no earlier analogy. There was no
history. The urges were to try to get warm, and try to stay fed.
In this latter role, you try eating this and try eating that,
very much as a small baby will do, if left to its own curiosity.
What is the average IQ at this time? Maybe 50 or so, in modern
terms. What was the sense of self? The "I-ness" was berely
coming out of the world of "me-ness." You were intellectually a
ten-year old. Can you go back in your own lives to the age of
eight or ten or so, to that pre-adolescent time of the magical
discovery of the "I" that took the place of the "me" of the
earlier years of childhood? The transformation of "He hurt me,"
to "I was hurt?" The heart of this experience has been described
in C.S. Lewis' autobiographical sketch, "Surprised by Joy."

And look at your involvement in this primitive scene. You
are physically superbly coordinated, and you are curious, but you
are maybe 25 years old, and you are considered by the others as
being an old man or woman. Most of your peers are no longer
alive -- they have disappeared into the whatever or wherever
that people disappear into. And whom do you ask to explain this
whatever or wherever? Someone else who has this knowledge. He
is the one who wears odd things that makes him different. He has
the same number of heads, arms, legs, but he is different because
he has the animal teeth slung around his neck, and he has the bag
at his waist with the magic things in it. He is the Shaman, the
medicine man, witch doctor -- later, the priest, the seer -- who
knows how to talk with the unseen spirits, knows what they want.
He knows the stories about the times of your ancestors, and
tells you what will happen in the future. And he knows how to
guide the tribe's actions in the present. All is, and will
remain (unbeknowst to you, of course), an oral tradition. There
is no written record. Oh, maybe a scratch here and a notch
there, as reminders of events, and of discoveries.

You have a soul (the term is our modern one) and everything
else has a soul. The bush, the stone, the child, the tree whose
leaves you eat, the animal whose leg you eat, all have souls.
But where are the bodies of pain and of sickness (wrongness) which
are things whose souls have invaded you? A tree hit by lightning

2

will fall, and that can be seen, but where does one look for the
cause of damage when a person has a toothache, or begins acting
strangely? Thus were demons created, spirits that couldn't be
seen. The demon that caused a toothache may have had huge teeth,
in the minds of early men; the earache demon possibly had
monstrous ears and talons. You couldn't see them but they had to
be there, since there were obviously things, spirits, that caused
illness and other disasters. How could a human being negotiate
with these unseen powers? There was one person in the
tribe -- the shaman or medicine man (sometimes a woman)-- who had
a way of communicating with these spirits. (It is interesting to
consider the possibility that a family group or tribe without
such a "wise one" might not have survived at all, in this
early world.)

If a person was possessed by a sickness demon, there were
certain noises and chants made by the shaman, certain steps and
movements of the body which could drive the spirit out. The use
of incantations, litanies and charms would work, as they still do
today for most of the people alive on earth. And the person who
was acting strangely was assumed to have lost his soul, perhaps
having had it stolen. His soul had to be found. The shaman or
wise woman knew what to do, and after demanding the necessary
ritual activities, perhaps penitence for sins, from either the
patient or his family, would find the hidden hole in a distant
tree where the missing soul was hidden, and pound it back into
the person who had lost it.

It is well to remember that, in our own time, many people
still believe that mental illness is caused by demon possession,
and that the victim must be assumed to have sinned seriously, in
order for this to have happened to him, and should therefore be
punished for his affliction.

The tribal wise person who knew how to find a missing soul or
drive out an illness demon probably made use of an altered state
of consciousness which made clear to him, in some form, certain
connections between lack of physical or mental balance and
particular states of mind and emotion. (Later, in Lecture 7,
we will spend an entire hour discussing nothing but consciousness
and its alteration without the use of drugs).

You are familiar with this kind of alteration; after all, you
have your own sleep-time and dream-world, and during your time in
the dream-state, your waking reality is just as non-real and
inaccessible as the sleep landscape is to your waking world.
But the Shaman can use this shift in consciousness to heal and to
seek out answers. You must recall that at this time in your
history, it was not the patient but the doctor who used the leaf
or the plant that enabled him to change his level of perception;
it gave him the insight into how to cure, how to find, how to
remedy, how to appeal to the person who had the tooth ache or the
broken bone, or who was about to die. The evolution into the
green medicine that treated the patient was many unseeable
generation ahead of you.

3

THE NATURE OF DRUGS

In the earliest days how would one treat a broken leg, from which blood gushed? You knew, from killing animals for food, that they were filled with blood (the knowledge of circulation is very modern) and that you, too, were filled with blood. And how do you save someone? With a hole in the blood bucket, blood pours out, so you stick your finger in the hole. But this is hardly a proceedure that can go on for days, so you find something, a large leaf, a piece of bark, a handful of grasses. Perhaps the break itself was wrapped in a special leafy splint that somehow made a stronger repair, and with time the bark that stopped the bleeding or the leaf that protected against infection or made the hurt go numb, was chosen over the grass or leaf that didn't.

The tribe that used a certain plant in their rituals had more and stronger children, and in time this plant was invested with the role of having the power of fertility and energy. Rather, it was the soul of this plant that assumed the form of fecundity and strength. And with the development of more stable groups, the change from hunters and gatherers out there, to the growing of the grains and berries right here, the bringing of food sources to the consumer and thus agriculture and the domestication of animals, came the parallel emphasis on plants that fed needs other than the tummy. This move to planting and harvesting called for attention to the seasons with its high sun and long days and its low sun and short days. And thus to the stars and planets that are moving continuously past the earth in some mysterious way. And the Shaman followed this, and predicted, and in his own way appeared to bring these things to be. It is a reasonable development then, from the Midwinterblut sacrifices which guarantee the return of a growing season, to the transferance of the unseen demons from the hard-to-fine hole in the side of the tree, to the less accessible skies themselves. And the demons became known as Gods, and the Shamans became known as priests. And the stars were the structures of the Zodiac, that embodied not only our food supplies, but our health and illnesses, our origins, and our fates. Astrology was truly an early science.

But how did the Shaman, himself, discover the plants that gave him these choices of consciousness states? For the most part, we cannot even guess. Some drug origins are easily surmised. The stumbling onto a bee hive amongst the rocks which had gotten wet in a rainstorm a few days earlier and had fermented, was surely the discovery of alcohol. The taming of fire and its maintenance with the wood of marijuana or the straw of the opium poppy, would produce smoke that caused dreams, and surely led to the discovery of these drugs. But the plants that were the sacred province of the Shaman, those that taught and guided, are almost without exception poisonous, inedible, and most often extremely hard to locate. It is almost as if the creation of these plant spirits, the investment of magic into the plant world, was done with the intent of making accidental discovery unlikely. After all, the knowledge of the identity and

4

location of such plants was a good deal of the shaman's power.

And all of this is surmise, of course, as there has been no available record until the last few thousand years or so.

But even in this day of record and tangible history, there are many examples of the exquisite union of pharmacology and the human consciousness that give believability to this sort of historical fiction. An example that I learned of, just a few years ago, was the procedure for killing a member of your tribe, in a culture where violence is totally taboo. There is a tribe of people in New Guinea called the Fore, which have been resisting the efforts of modern man to civilize them, as they are believed to be cannibals. They are, in fact, vegetarians and totally incapable of violence of any kind, either against themselves or against any other living thing. They will apologize to the plant while gathering it for food. And yet, part of their payment of respect to a recently died member of their family, is to ceremoniously eat part of the physical remains of the body. But in this society, as in every society, there are taboos of such a treatening nature, that the violator must be removed from the tribe. Better lose one than lose all. How does a non-violent tribe kill an offender? Perhaps banish to the bush. There he will either starve, or be killed by some neighboring unfriendly tribe. Certainly not this tribe's own problem anymore. But some taboos require tangible punishment for its deterrent value.

The answer is to be found in that overlap of pharmacology and psychology. A short piece of bamboo, about three inches long and an inch in diameter, which contained a node in its center, was sharpened on one end to a fine point. At the other end, the hollow center was filled with a mixture of several things, among which were the ashes of a certain plant. Then during the night, this instrument of execution was driven into the ground directly in front of the home of the condemned. He would arise in the morning, step outside and see the symbol, and recognize its message. He would return to his dwelling and inevitably he would be discovered dead within 24 hours. He knew that he had been found guilty of a capital offence against the tribe, and would choose to die. No violence, not even by himself against himself. He had been raised to know that this is part of his reality. His death was a voluntary yielding of life.

We have all have similar convictions as to our roles within our society, and to the Shamans-priests-physicians to whom we turn for advice and help. When you are hurt, physically, and appear at the nearest hospital, you turn to the the white jacket, the stethescope, and the name-tag with M.D. on it. If the helper were in dirty jeans, wearing an old shirt with the tail hanging out, and shod in skuffy sneakers, well, he simply cannot be your helper, somehow. His medicine won't work as well as that from the proper image. He doesn't seem to have the authority you seek. And if the hurt is spiritual, you turn to the cloak and symbol of your personal faith. Here, too, it is unlikely that

5

dirty jeans would do it. Our Shamans must always stand apart
from us, else how would we know them?

But back to the story of the history of drugs. Into
recorded times, one can find still the embodiment of health, and
life, and fear and death in the animal Gods of Egypt, and into
Greek and Roman times, this was the device for curing and that
was the device for poisoning. The casting of the entrails of a
chicken, or the fall of the magic sticks of the I Ching, could
help in the diagnosis of a malady, and a few fundamental plants
were ubiquitously recognized for their dramatic influence on man.
But none of these were seen as instruments in their own right for
the treatment of illnesses. They provoked, or comforted, or
instructed, but the illness demon within the sick person still
had to be exorcised as a demon, to allow the body to become
unpossessed again.

Even up to the almost immediately recent time, the making of
a connection between agents such as bacteria and the illness, and
the concept of using a drug to cure, were unknown. Microbes were
seen and studied 300 years ago, but their association with
illness is barely a hundred years old. The concept of
chemotherapy, and the designing of drugs to destroy pathogens,
are products of this century. The term "Pharmacology" was
created in about 1890. Before that, the medical schools spoke of
the discipline as Materia Medica, the stuff of medicine.
Surgeons until recently addressed the patient wearing formal
dress, with tails and hats. Infections came from the air. The
very disease Malaria itself is named from the French, bad air,
from which it arose. You always became ill with it in the
swamps, after all. The use of the Cinchona Bark for the relief
of Malaria was a revolution in the mid-1600's in that this was
one of the first awarenesses that sicknesses need not inevitably
be treated through the voiding of the abnormal bodily fluids,
through blood-letting, vomiting, and purging. And yet it was
almost two hundred years before the concept that plants were not
homogeneous things but actually contained individual components,
and another 50 years, into the latter part of the 19th century,
that Quinine was isolated and found to influence a plasmodium
that had been associated with the disease.

The body of understanding that gave birth to modern drugs,
was developed in the middle ages under the term Alchemy. The
popular myth concerning alchemy is that its goal was to transmute
base metal (lead) to gold. This accounts for the aura of secrecy
and diligence that has been ascribed to alchemists, and sadly
belittles their intent. The transmutaion that was being sought
was in themselves, and in observing the fundamental changes that
can be wrought in the very foundations of the real world by the
simple application of heating, cooling, dissolution and
extraction, then they too, as curious and learning individuals,
could be transformed as well. From the sick to the well? From
the weak to the powerful? From the mortal to the immortal? Are
we so different today with our efforts to live forever by leaving
behind us our authored books, our created children, our records

6

of good and bad deeds?

What were these foundations of the real world? There were four in total. The world was divided in dry and damp, and into warm and cold. All real things were believed to fit into these categories, as were all of human characteristics as well.

FIRE △ or ○

The dry and warm was the element of fire. The terms fiery and choleric stem from this, and the human characteristic was that of being hot-tempered, impatient, and easily irritated.

AIR △ or ☉

The damp and warm was the element of air. The association with blood is classical, and the personal attributes are those of being airy, sanguine, and hence, sturdy, high-colored, confident, and optimistic.

EARTH ▽ or ⊕

The dry and cold was the element of earth. This was the solid and melancholic character, with the overtones of depression and of sadness.

WATER ▽ or ⊖

The damp and cold was the fourth element, water. This is the fluid and phlegmatic aspect of man, and accounted for sluggishness. The person was stolid, unemotional, and impassive.

And it was with these concepts that the alchemists worked. To manipulate all things about you, from the minerals of earth to the menstrual fluids of mankind, with the four basic elements of heating, cooling, wetting, and drying, was believed to allow in time a complete understanding of matter, and of course one's self. The search was always alert to the fifth element that was without material reality and that Aristotle assumed permeated the whole world. The concept lives with us today in the term "quintessence", the fifth essence that represents the spirit and soul of something. It was toward these goals that the Alchemists labored, and if they could make gold out of lead, it would be to find a form of gold that could be drunk. The name of this magic stuff was "Aurum potabile" and it was conceived of as an elixer, truly the Philosopher's Stone, that would thus be the true tree of life. It would gratify all desires of the person possessing it. It would rejuvenate, retard old age, strengthen, and restore hair. And if properly applied, it would actually prevent hair from turning grey. Much of the supposed elixer was sold in the market place, but little of it, if any, was ever actually made.

And there were the elements of the Ancients. There were seven classical elements, and they corresponded in their value, and rareness, to the seven moving objects in the sky. The fixed

7

backdrop of stars was held in magical contrast to the seven
moving objects which also varied in their importance depending on
their brilliance. And the names of the days of the week also
took their origins from this structure. From the brightest, most
valuable, and most important, to the least, these were:

Element:	Sign:	Planet:	Day of the Week:
Gold	☉	Sun	Sunday (dies solis)
Silver	☽ ☾	Moon	Monday (dies lunae)
Lead	♄ ♄	Saturn	Saturday (dies saturni)
Iron	♂	Mars	Tuesday (dies martis)
Copper	♀	Venus	Friday (dies veneris)
Tin	☿	Hermes (Mercury)	Wednesday (dies mercurii)
Electrum	♃	Jupiter	Thursday (dies jovis)

These elements were spun into the vocabulary of the
Alchemists in every way. Gold was Aurum in Latin, but was
considered quite noble and immutable. Silver can be referred to
even today by ancient names (Lunar Caustic for Silver Nitrate, a
corrosive known for centuries to protect against disease without
knowning what a disease even was); Lead poisoning is still called
Saturnism; Mars is the God of war, as the element Iron is even
yet the instrument of war, and both Copper and Venus still
represent the female. Electrum is now known to be an alloy of
gold and silver, but was then thought to be a single substance.
Tin can be found here in its place, on occasion. To this list
the alchemists added literally hundreds of new elements,
metaloids, and novel chemicals.

Signs of the other planets:

Signs	Planet
♅ ♅ ♅	Uranus
♆ ♆ ♆	Neptune
♇	Pluto
⊕ ♁	Earth
⚦	Vulcan

8

One sometimes gets a strange feeling that there might be an interchange between cause and effect in this remarkable listing. What is seen here might not necessarily the happenstance of effect, but might have some component of causality. If a person interacts with something and gives it character, this character then may well become imbedded in that something, and will be accepted by others as an intrinsic property of that something.

It was from this vast body of observations that chemistry evolved. There were many attractive but false leads. Neither the vital fluid that allows life, nor the phlogiston that allows fire, have ever been isolated in a bottle. But other leads were well ahead of their time. The first preparation of diethyl ether occurred in the 1200's by the treatment of alcohol (from wine) with sulfuric acid (Oil of Vitriol, from burning sulfur). It was called Sweet Vitriol, as it was known to taste sweet. But it was 300 years later that the remarkable Physician and true father of chemotherapy, Paracelsus, introduced it (as Sulphure Embryonato) to an animal. He found that if fed to chickens, it put them to sleep. Yet another 300 years would pass before it was taken by man, and then largely as a party intoxicant. Its use in medicine as an anesthetic followed much later.

But these concepts of chemicals, minerals and alchemy paved the way for our modern world of drugs as surely as did the plants of antiquity. The concept of chemotherapy is taken for granted now, but the concept is less than a hundred years old. The idea of designing a specific compound to be more toxic to an unwanted guest in the human body than to the human himself originated with Paul Ehrlich with the synthesis of Salvarson, the famous 606th compound in a study aimed at the killing of the Syphilis spirochete without killing the infected host. This was, believe it or not, in 1907. The concept of drugs as such did not exist much before this, since the concept of illness was related to completely intangible origins. Something you had done, or somewhere you had been, or something should have said but hadn't (or shouldn't have but had).

The Plague that struck Europe 1347 (the Black Death, probably the major factor in bringing the Dark Ages to a close, as it struck all, from rich to poor, effectively destroying the feudal structure), was believed to be so contagious that merely to see a victim was to become a victim. Physicians who tended the dying, themselves died. But from it came new practices of medicine, such as the quarantine (40 days, because it was known that all acute diseases, like the biblical flood, were limited to that time period), and for the first time the airing of homes and the burning of suspect clothing supplanted the classic reliance on prayer and resignation. And when the Plague returned in the 15th century it was far less destructive. The London Great Plague of 1665 was probably aborted by the sweeping fire of the following year. But even though the spread of the Bubonic Plague was early on associated with the spread of rats, no causal connection was ever suggested. The rats were an omen, nothing more.

9

God gave diseases, and God took them away. Cleanliness was felt by some to aid the medical profession, but to others, certain kinds of uncleanliness were equally strong proofs of success. The wrapping of the bloodied rag outside the blood-letter's office is still seen today as the barber pole. By statistical logic, the concept of cleanliness eventually prevailed, but one must remember that the practice of medicine has never been, nor will ever be, a statistical art. It is basically a one-on-one relationship. And statistics do not apply to individual cases.

This can be seen dramatically today in the opposite philosophies shown in the training for an M.D. degree or for a Ph.D. degree in the medical sciences. In the preparation for the M.D. degree, the argument is maintained that the primary goal is to cure the patient, and if some generality should come from the interaction, that is all to the good. In the preparation for the Ph.D. degree, the opposite argument obtains. The primary goal is to discover generalities, and if some value can be applied to a specific case, then that is all to the good. The numbers of scientists that share both degrees, and are at peace with the intrinsic contradictions, are very few indeed.

And although the concept of selective toxicity (chemotherapy) did not exist before the 1900's, it could be applied only to protozoan infections. The application of this concept to bacteria, much simpler and unsophisticated organisms, had to wait another quarter of a century. A red dye, Prontosil, was discovered in 1932 to be effective against streptococci infections in a living animal by Gerhard Domagk, but it was not active when it was administered directly to the microorganism. It needed the host animal as an intermediate. In man, the compound is broken into two halves, one of which is the active antibiotic, Sulfanilamide. Now the art of the chemist took over, and many dozens of variously modified Sulfanilimide structures were made, and found to have a broad collection of specificities.

At about this time, the art of the microorganism itself as a synthesizer of antibiotics was discovered. Penicillin was discovered to have antibacterial properties, and was first used in man at the time of World War II. This gave rise not only to many synthetic variations, but to the search for other microorganisms which might also have antibiotic properties. The Streptomyces group proved especially rewarding, and drugs such as Erythromycin, Lincomycin, the toxic but extremely effective pair, Vancomycin and Novobiocin, soon followed. Yet others known as the tetracyclines and the rather simple drug Chloramphenicol, are effective not only against most bacteria but against parasites such as the rickettsias. Hence the term, broad-spectrum or polyvalent antibiotics.

Remember, the whole concept of drugs and drug use in the treatment of disease is less than a 100 years old. Which is, by no coincidence whatsoever, almost exactly the period covered by

10

the writing of laws that have been designed to regulate, and control, the use of drugs.

HISTORY OF THE DRUG LAWS.

The first laws in the United States directed towards efforts to regulate and control drugs were written just before the turn of the century.

Historically, however, the proscriptions against the use of plants that gave pleasure or insight are as old as the use of the plants themselves. If the elite minority who has been given the power to dictate social policy is the group that itself finds benefit from the drug (plant), then social policy (drug law) is established to prohibit its use by the non-elite majority. If this elite minority who can dictate social policy does not use the drug (plant) and thus itself receives no benefit from it, its use by the non-elite majority is seen as a threat to its authority, and social policy (drug law) is established to prohibit its use by the non-elite majority. And if the elite minority shares with the non-elite majority the drug (plant) in question, no social policy (drug law) is written since the substance in use is not called a drug.

The first example is seen in the ritual restriction on the use, by the Shamans and medicine men of yore, of sacred plants such as Peyote and Teonanacatl. The second example is that of the influence of royalty in matters of the introduction of foreign drugs (plants) into the society that was under their command. Examples are the introduction of tobacco and coffee into Europe. Their use was, in some instances, punishable by the death penalty. And the last example is that of Congress here in the United States, who share coffee and alcohol and tobacco with the common man, hence they are not considered as drugs.

But woe be it to a minority that introduces a new drug into the culture. Although the origins of Opium were probably in middle Europe, its cultivation and use spread Eastwards, in almost the exact mirror image of the Westward spread of the cultivation of Hemp, from China through India, to the Arabic-speaking cultures of the Near East. The use of Opium in China became widespread, largely as an outgrowth of the Opium Wars of the early 1840's wherein the British forced the Chinese to accept Opium in place of gold in payment for the wealth of Chinese exports that were much in demand in Europe. I will get into more detail on this sordid chapter in the Lecture on the Narcotics (Lecture # 12). Opium was a familiar and favorite drug of the Chinese coolies, who were imported into this Country as cheap labor during the 1860's to work on the Tran-continental railroad. But once the last spike had been driven, the labor force was seen to represent a threat to American labor, and there was a strong effort made to get them to go home.

11

A San Francisco Ordinance in 1975 forbade the keeping of Opium Dens, and such prohibitions were passed in turn by one Western state after another. The Federal Congress passed a law in 1883 raising the tariff on smoking Opium, another in 1887 prohibiting its importation into the United States by any subject of China (an unrecognized bit of blatant racism), and finally in 1909 a law prohibiting its importation my anyone. Actually, Morphine (the active ingredient of Opium) was already widely used in the United States, having been a mainstay for pain control during the Civil War. It was at this time that the hypodermic needle (the hollow needle) and the syringe needed to push a fluid through it, were made widely available. In the post-Civil War years the use of morphine spread widely through the country for the treatment of almost everything. Some of the claimed virtues were valid, but many more were out-and-out fraudulent misrepresentation. Most of the promotion was directed towards women and their complaints, and dependency upon it became known as the "Women's Disease."

But morphine was swallowed, never smoked. Smoking of drugs was a dirty, somehow evil and un-American thing to do (remember the connotations of the Opium Den?) and it was done with Opium by a visible and unwanted minority. Smoking of Marijuana was done by an invisable (and unthreatening) minority and so was ignored. And the smoking of Tobacco was done by everyone (at least if they were of the male sex) and so it was not considered to be a drug. The major effects of these legal restrictions appears to have been two-fold: there was the shift from Opium smoking to the use of still legal Morphine, and the creation of an illegal sub-class of Opium smokers who chose to continue their preferred habit.

It is noteworthy that the concept of smoking is not mentioned in the Bible, and was therefore contrary to the teachings of strict Biblical constructionists. The art of smoking was uniquely a New World phenomena, being unknown in all of Europe, Africa, and Asia. When the early explorers from Europe met the American Indians, they were amazed to see Caribs approach them with ignited leaves stuck up one nostril. They took the plant (Tobacco) and the technique (by mouth rather than by nose) back to Europe with them, and it was there that the procedure was elaborated to embrace Hashish and Opium. The smoking process soon swept around the rest of the World.

One of the concessions of the Spanish-American War of 1898 was allowing the United States annexation of the Philippines. This was the first Asian colony for the United States, and with it came a large social and health problem associated with the use of Opium. The concerns concerning Opium use, both domestically and abroad, caused the United States to play a major role in the Hague Convention of 1912, which outlawed International Opium traffic which was not for medical use. And at about this time, the broad use of Heroin was becoming a concern. It had been introduced in medical practice in 1898 as a pain killer and as a treatment of Morphine dependency, but it was soon recognized as producing a similar dependancy.

12

It became an embarrassment to the United States that it had
no domestic legislative policy concerning drugs and drug use.
The Harrison Act was passed in 1914 which illegalized the
importation, sale or possession of either Opiates or Cocaine
except within medical channels. This Act was to remain, with its
many amendments, the central legal agent concerning drug policy
for over 55 years.

Two aspects of the law warrant special comment. The
enforcement foundation of the law was based on fiscal matters
rather than on criminal aspects. The agency responsible for its
administration was the Department of the Treasury, rather than
the Department of Justice. All legal manipulations concerning
drugs required licences and the payment of fees, and the criminal
act was the failure to obtain these permits.

A second aspect concerned an unforseen fine line that would
have to be drawn between the medical and the criminal aspects of
drug addiction. There has always been an understanding between
the legislative and the medical communities, that no law shall
ever be enacted that shall dictate the practice of medicine. In
the final text of the law, the following phrase appeared:

"Nothing contianed in [this act] shall apply to the
dispensing or distribution of any of [these] drugs by a
physician ... in the course of his professional practice
only."

It is difficult to predict what a simple sentence, when
spoken by a determined person in a position of power, can mean.

The Harrison Narcotics Act had been debated in Congress for
several days, but never was there mentioned any problem of
narcotics addiction in the United States. None was recognized.
Everyone spoke about the 1912 Hague Convention involving
international obligations. There were no concerns about domestic
morality. And the bill itself was not one of prohibition. It
was to assure registration of producers, importers, and
physicians who prescribe narcotics, by the collection of a modest
tax. Most over-the-counter preparations were exempted as long as
they contained less than two grains of opium, a quarter grain of
morphine, or an eighth grain of heroin. And the physician's
rights were explicitly protected. They were required only to
keep records of prescribed drugs.

The target of the anti-drug enforcement crusaders was the
phrase: "In the course of his professional practice only."

Within two years the law enforcement officers decreed that
narcotic addicts were criminals, not patients. Addiction was a
crime, not a disease. And thus any prescription in this area by
a physician was clearly outside the course of his professional
practice. Many physicians were arrested, and some were
imprisoned. And a law designed to regulate started the smooth

13

transition to one that espoused prohibition.

In the telling of the history of drugs, a story that is parallel to the drug abuse aspects (the Harrison Act) is one directed to drug use. A bill entitled "The Pure Food and Drug Act" was passed in 1906, and was placed under the administration of the Department of Agriculture. Its purpose was to require the accurate labeling of the vast number of patent medicines that were being sold in the Over The Counter (OTC) market, but there was no effective way that it could be enforced.

Another anti-drug chapter was written back in this post-World War I era; the passage of the 18th Amendment to the Constitution prohibiting the manufacture, transportation, importation, exportation, or sale of alcoholic beverages. There was an Alcohol Unit established in the Department of Treasury, charged not only with the enforcement of the prohibition amendment, but with the anti-opiate and anti-cocaine portions of the Harrison Narcotics Act. This Constitutional amendment remained in effect until its repeal in 1933 that allowed weak beer to be sold, and then hard liquor the following year.

The absence of an enforcement groups dedicated to the Narcotics and Food and Drug laws, was felt to undermine the strengths of both laws. This was corrected by the establishment of the Food and Drug Administration (FDA) in 1927, and the Federal Bureau of Narcotics (FBN) in 1930.

Only one major change was made in the Harrison Narcotics Act between then and the period of chaos and confusion that occurred during the 1965-1970 period. The repeal of the prohibition amendment transferred back to the newly formed FBN the enforcement responsibilities dealing with the Opiates and Cocaine, and with this transfer came the former Assistant Prohibition Commissioner Harry J. Anslinger who took over as Commissioner of the FBN. He saw the use of some new drug problem as a vehicle for the promotion of the image and the power of the Bureau, and started a campaign against the lethal weed known as Marijuana. This resulted in the passage of the Marijuana Tax Act in 1937, which placed Marijuana into the Harrison Narcotics Act as a narcotic drug. Again, the proscription was a fiscal measure designed to control by the imposition of licensing and taxation procedures.

There was during this time, however, a number of amendments to the Pure Food and Drug Act, attempting to broaden and strengthen its influence on and control over drugs intended for medical use. The Food, Drug and Cosmetic Act of 1938 introduced requirements that drugs be shown to be safe before being allowed for medical use; another amendment in 1951 established the structure of the Prescriptional Status for drugs that were deemed to be sufficiently dangerous to the patient that a registered physician should oversee their usage. And another amendment in 1960 added the additional requirement that, before a drug can be sold, it be shown to be efficacious in producing the virtue

14

claimed for it.

The period between 1965 and 1970 was one of social revolution in this country, with the increasing rebellion against the Vietnamese War, and the seemingly indifferent attitude of the Johnson Adminsitration towards domestic policies of desegregation and civil rights. One of the expressions of this unrest was the spawning of the "Psychedelic Revolution" with its emphasis on individual rights. This was expressed by a rapidly broadening pattern of drug use and drug abuse. This aspect of the protest was seen by Congress as a deliberate flaunting of authority in the area of the Drug Laws. Many changes were rapidly instituted, and it took several years for a new structure to stabilize.

The passage of the Drug Abuse Control Amendments in 1965 (to the Harrison Narcotics Act) removed all non-narcotic drugs from the authority of the BND and placed them under the control of a new agency, the Bureau of Drug Abuse Control (BDAC), to be administered by the FDA. The narcotics (Opium, Cocaine, Marijuana (!)) stayed under the Department of Treasury, and the stimulants and depressants were now the province of the FDA. During this short period, a number of the Psychedelic (Hallucinogenic) drugs came to official attention, and they fell to the FDA organization, first as Stimulants and then in a category called Hallucinogens. The FDA had never been an arm of criminal law enforcement before, and there was quite a bit of confusion and uncertainty of authority delegation between them, and the FBN (and Customs, as well). This was resolved in 1968 by the amalgamation, by President Johnson, of the BDAC and the FBN into a new agency, called the Bureau of Narcotics and Dangerous Drugs (BNDD) to be administered under the Department of Justice finally.

The transition was made more confused by the passage of an entirely new Federal Drug Law in 1970, known as the Controlled Substances Act, bring to a close the era of the Harrison Narcotics Act which had been the guiding principle of anti-drug law for over 55 years. Several supporting and advising drug-enforcement groups came into being over the next two years, mostly under the Department of Justice. The major ones were the Office of Drug Abuse Law Enforcement (ODALE), the Office of National Narcotics Intelligence (ONNI), and the Intelligence and Enforcement functions of the U.S. Customs Service. Nixon issued a Presential Order (in 1973) that brought all of these separate agencies into one new group, called the Drug Enforcement Administration (DEA) which is still the primary enforcement arm today.

The 1970 Controlled Substances Law instituted a new concept, the prioritization of enforcement attention and of criminal penalties based on the abuse potential, the potential hazard, and the recognized medical utility of the drug that is made illegal. There were five Schedules established, ranging from Schedule I (a drug with a high abuse potential and no recognized medical utility, to Schedule V (a drug with a low abuse potential and a

15

recognized medical utility).

The legal definitions of these Schedules is instructive.

SCHEDULE	ABUSE POTENTIAL	ACCEPTED MEDICAL USE?	RISK	
I	High	No	There is no accepted safety under medical supervision	
II	High	Yes	What is the dependency risk that will follow abuse?	
			Psychological:	Physical:
			Severe	Severe
III	Less than with I & II	Yes	High	Low or Moderate
IV	Less than III	Yes	Limited (Less than III)	Limited (Less than III)
V	Less than IV	Yes	Limited (Less than IV)	Limited (Less than IV)

The procedure needed to place a new drug within this structure involves a number of steps. An announcement of intent must be published in the Federal Register. A waiting period of 60 days allows comments and objections to be received. If it is felt to be appropriate, hearings will be held to address these matters. Otherwise, a final action can be taken at the end of the sixty day period. The recently passed Comprehensive Crime Control Act of 1984 allows the bypassing of this delay in instances of drugs that are felt to present an especially serious threat to public health. A temporary emergency scheduling may be invoked for a period of one year, to apply during the period allowed for comments and hearings, and it can be renewed for an additional 6 months following that first year. Action must be taken before the expiration of the renewal period.

Schedules I and II are the categories with drama and attention-grabbing names. Here are the forbidden drugs (Schedule I) such as LSD, Heroin, Marijuana and Peyote) and those which are equally newsworthy but for which there exists some medical use (Schedule II) such as Methamphetamine and Cocaine. The sleeping

16

pills and minor tranquilizers, drugs with less popular impact, are found in Schedules III to V.

Problems have arisen with this scheduling structure. In the matter of vocabulary, there is no definition of what is meant by either "currently accepted medical use" or "high abuse potential." There are four levels of abuse potential explicit in the law, and no suggestion as to how to obtain a quantitative number for this hazard. Both expressions have been linked to the presence or absence of approval for use (or exemption from such approval) as proclaimed by the FDA. The medical community does not agree. And from the listing given above, there is no Schedule in which to place a drug that has no accepted medical use but which has something less than a high abuse potential. These problems have been brought into focus with the recent Scheduling of the drug, MDMA.

From about 1976 to date, there has been approximately one law passed per year that bears on the legal aspects of drug control. Some of these increase penalties, others broaden authority or subvert rights, all for the noble goal of bringing the drug-abuse problem under control. Some of the more influencial of these are:

In 1978, the Psychotropic Substances Act allows for the first time the application of criminal forfeiture allowing immediate seizure of assets and property believed to have resulted from drug-related criminal acts. This is the first allowance of criminal forfeiture since it was declared to be inappropriate by the Founding Fathers in 1780. This law also allows the International Convention on Psychotropic Substances the privilage of suggestion additions to the Controlled Substances Act.

In 1982, the Department of Defense Authorization Act reestablished the Posse Comitatus, allowing for the first time the involvement of the military services in the enforcement of civil law, if it involved drug-related crime.

Also, in 1982, the Tax Equity and Fiscal Responsibility Law removed restrictions on the tax record availability in narcotics prosecution cases.

In 1984, the Comprehensive Crime Control Act (mentioned above) allowed emergency scheduling which bypasses the usual safeguards of public hearings.

In late 1986, just before the Congressional Elections, a broad law was passed that effectively outlawed any human research with any drug that had stimulant, depressant or hallucinogenic effects, unless that research had been explicitly approved by the FDA. Any manipulation with such drugs (possession, intent to give to man, etc.) will be a felony and punishable as if the drug was actually a Schedule I or Schedule II drug, even though it need never be proposed for Scheduling. A structure that is

17

substantially similar to a Schedule I or II drug is sufficient to condemn.

A comment may be justified in connection with the strange term "substantially similar." There is a term used in rhetoric, called a disclaimer. It is a loop-hole in an otherwise firm statement. Words such as "almost", "pretty much", "about", by-and-large", "more-or-less", "approximately" are all disclaimers in that they give you an escape from a rigid statement. I worked for a couple of years with a master multi-disclaimer. His masterpiece was, "We will probably ship the sample to you in 7 to 10 days, or two weeks at the latest." Three disclaimers in a single sentence. Here the law uses two such words in a single phrase, leaving the reader unsure of the meaning. "A structure that is similar to another" is the same as "A structure that is substantially the same as another." But what additional freedoms of interpretation are implied by the phrase: "A structure that is substantially similar to another?" You have the feeling that the words were put in there to provide whatever latitude might be needed in the enforcement structure.

Each of these legal changes has, in its time, been lauded by the law enforcement community as having proven to be an effective weapon. But each of them has removed another freedom from the American public, and there is no objective evidence that the drug problem is diminishing.

A final note is appropriate, relating to the progression of the names for the drug enforcement agency empowered to administer Federal drug laws. We have seen the transition from FBN to BNDD to the present DEA. A move is now well underway that is effectively transfering all administrative authority from the DEA to the FBI, and it seems clear that within another year, the FBI will be the regulatory body controlling the enforcement of the drug laws.

A brief review of the major dates and events in the history of the Federal Drug Laws is attached. It was handed out in class at the start of Lecture # 4. It is taken from "The Controlled Substances Act, A Resource Manual of the Current Status of the Federal Drug Laws", by A. T. Shulgin, January 1, 1987.

18

One first felt the sudden silence. With the reverent anticipation of being a listener before a symphony, we awaited the appearance of the greatest medicinal chemist of his age: a rogue, eyes alight, white hair flowing, electrical emanations from every gesture. He had a true love of chemistry. There will be no other like him.

The privilege of being taught by Sasha was, from the moment we met, the transformative event in my then limited life. I had failed to solve a problem in advanced chemistry. Only one man was capable of the insight, and was of like mind in advancing the procedure: Sasha. It was 1983 when I ventured to write him of my long admiration. of reading his many papers, and of the difficulty in an esoteric realm of synthesis. Sasha replied with an invitation to join his lectures in drug studies at San Francisco State University. I hastened to attend: honored, timid, excited.

During Sasha's lectures I smiled with delight, for here was one devoted to the ancient art, joyous at the interplay of molecules and atoms, one who envisioned the future of medicine, whose chalk flew across the last blackboards with elegant structures of his beloved phenethyamines and tryptamines and the many analogues he invented. To entice, humor, and comfort undergraduates frightened at the mythic difficulty of organic chemistry, he referred to his drawings as "dirty pictures." At this, students often laughed, knowing he was playful and understanding, one who saw the beauty of this mysterious, sub-microscopic world. By then, he had us all.

His loving Ann Shulgin always attended the SFSU lectures, notepad in hand, though she was familiar with the stories of wonder. Their daughter Wendy, with her long blond hair to her waist, often was present as well, fresh from travels in Asia or the Middle East, and much to the distraction of the young men devoted to Sasha. Those present sensed the beginning of a new world, for we were among the privileged who bore witness to the anecdotes, observations and teachings of a modern shaman. We became enthralled at the mysteries laid before us.

In time, I would remember this unparalleled, gentle, splendid man in *The Rose of Paracelsus* ....

> Everyone must get to experience a profound state like this.
> I feel totally peaceful.
> I have lived all of my life to get here, and I feel I have come home.
> I am complete.
>
> — Experimental journals of Alexander Theodore Shulgin, PhD

ONE PASSED an old, dented orange Volkswagen, a Kharmann Ghia of better days, bearing the California license plate OSOMO—the Japanese for peace. From their artists' abode of rambling structures came Ann and Sasha Shulgin, elders and aristocrats of the spirit. Welcoming their many visitors with embraces, they guided them within, even while a small outbuilding contained the most productive, licensed psychedelic research laboratory on earth.

A painting of Sasha in his boyhood—dressed as the "Blue Boy" and holding his treasured viola—occupied an alcove. A piano crowded the rough living room, strewn with cushions and papers and books. Their hand-built residence on a mountaintop retained the most magnificent view of the entire Bay Area, with the hillside spilling down to a plain by the sea.

Bric-a-brac stuffed a tumbledown garage. Charming little patios of cracked cement had shaded umbrellas and wooden tables, while a narrow, flowered path led to the small, almost ramshackle but utterly refined laboratory where new worlds were dreamed.

In Sasha's study were all known references on entheogens, little gifts from friends and colleagues, and a plaque of appreciation from DEA for his scientific expertise on controlled substances. On a bookshelf was a jeweler's brass belt buckle of the Grateful Dead alembic—a skull split by a lightning bolt—labeled "No. 1" and signed by its creator: the legendary underground acid chemist Augustus Owsley Stanley III. Among Sasha's multiple computer terminals and thousands of books and files and memorabilia, one became lulled by the water music of pure thought and affection, and the companionable silence of games of destiny.

There was only one bedroom, for the other contained a massive, derelict Nuclear Magnetic Resonance machine for structural analysis of experimental substances. The small kitchen was in eternal disarray. Despite Ann's best ministrations, she often was overwhelmed by streams of convivial visitors. Old sage picked from a field rested on a faded chest of drawers. It was a peaceful place, a sanctuary. Here one could see clearly, so that to our no longer flawed vision all was sun-golden and leisurely, like a conservatory smothered in rose creepers.

A LEAN, six-foot-four mass of wizardly white hair and deeply welcoming bear hugs, Sasha loved word play and humor, science and music. The ultimate authority on psychedelic chemistry, he engaged all the world's primary researchers in the field, from the president of Aldrich laboratories—the chemists' chemists—to the pharmaceutical department of Sandoz in Basel, and those few with very long memories of how it all began.

Each month, a Friday Night Dinner or "FND" was held at a friend's house, with Sasha and Ann always present. They often hosted small dinners, or picnics on Easter and July 4th weekends at their home. Small children darted about fresh as roses, finding favor everywhere, like little birds among cherries and raspberries. The gatherings were melting townships, for airs of nervous preoccupations and sudden shynesses dissolved as ebullient Sasha sketched in the air his alchemies and divinations.

With their natural affability and the quality of their science, the compound was a magnet for artists, psychiatrists, neurologists, engineers, musicians, faculty from Berkeley, Stanford and UCSF, philosophers, a spectrum of seekers, underground figures, scientists of every specialty, dancing girls, Edgewood Arsenal alumni, Silicon Valley magnates, poets, forensic chemists, code warriors and writers. All were grateful something so fantastic existed. This blending of hermetic schools was synergistic; we were the moon people again those nights, the cloud-forged and blessed.

Their kindliness hardly reflected Sasha's 300 rigorous scientific papers, patents and books on the chemistry and pharmacology of entheogens he had invented. He was lively, with a theatrical note and rapid, soft, playful speech, his mane of white hair and beard in chaste silver points. Everyone confided in Ann, for she was receptive, understanding, with a slightly trailing and sophisticated trace of accent from her travels abroad. Exerting her social magic among so many, she sometimes would blush a damask rose. Ann was Earth Mother to a world tribe.

The pervading and disciplined science, with accomplished professionals all around, was not austere or forbidding. Those who were unaffiliated with academia, or not yet enfranchised into research, were welcomed by scholars who were experienced in personal transformations. Some people arrived as honorary attaches from unspeakable realms, and now were invited to rest here in the cool bosom of new consciousness. For others, appearing from the covert worlds and fearful of groups, it was like coming in from the cold. A special few returned from the far mountains and secret deserts, from their endless loneliness and practice of clandestine arts. To these, Sasha and Ann seemed angels of light.

Laughter was everywhere, serious asides here and there and—penetrating everything—the everlasting sunshine of simple compassion. Poignant, ghost-like presences sometimes were apparent: the deceased and luminous forebears of these arts, the remembered and loved, the imprisoned, and around the edges always the lost, or those too fearful to appear.

One entered into superlative esoteric states near Sasha and Ann, for one was in the company of like minds at last. With studied politeness, newcomers first stood in the violet shadows, while Sasha with his jilted Anglo-Saxon launched into abstractions and twinkled with kindly complicity, inevitably drawing out his fond listeners.

Over time, in the growing acceptance and friendship all around, the gatherings had a dream detachment—one's perception changed as if by evening river winds—so that in the brindled autumn and summer moonlight all was safe. We were home.

IN MY 20TH YEAR, visiting the stacks at Mallinckrodt laboratories at Harvard, I was reviewing articles in the *Journal of Organic Chemistry*. Through a Gothic window, as sunlight fell upon the page, I first became aware of Sasha's work. Although following forever after all of his papers, I—only after quite some years—braved writing him. Suddenly I was an invited, eager student in forensic chemistry at SFSU and Berkeley, scribbling notes frantically as he lectured to packed classrooms.

By first announcing to his crowded classes that he would divulge some "dirty pictures," Sasha would introduce students to seemingly impenetrable molecular structures, precisely and rapidly drawn in chalk. It was not a reference to the licentious, of course, but his humor about the fearsome intricacies of organic chemistry so alarming to those with no background in the art. With his little joke, everyone relaxed. We were carried skyward, for everything became simple in his hands. Sasha rendered even the most arcane concepts elegant, even beautiful in a way.

To young people fascinated with the advances in chemistry and pharmacology, and who showed persistence in their talents, Sasha always pointed out the path of advanced degrees, guiding them away from the uncertain and vulnerable covert life.

"Get that PhD!"

I sought refuge with them when I could. During my first dinners at their home, alone with them in the years before they wrote the seminal *PIHKAL* (Phenethylamines I have Known and Loved), Sasha brought out his voluminous handwritten notebooks. We sat outdoors, under a black sky, with its plates of brightness and a surprising moon. Sasha and Ann had eyes of unparalleled depth those evenings, the night flowing down to the white earth. Within his decades of records were drawings of molecular structures, calculations, elaborate syntheses he had created, and comments on subjective effects of hundreds of new psychoactive molecules. Exploring the journals with reverence, I teased him lightly.

"Where's the page that says 'Eureka'?"

Laughing, he pointed out the structures 2-CB and 2-CE, then described their teachings, eroticism and aesthetics. Of the mysterious Aleph-1, he fretted about mania, or ego-inflation.

Striding about the Berkeley campus or at conferences from Amsterdam to Heidelberg, Sasha was easily spotted, so tall he was, with his mane a wreath of white flowers. His personal vivacity was constant, for he possessed an innate joy. He would appear as suddenly as a star slides into the sky.

HIS LABORATORY was out back. In the warm twilights, under a curtain of wisteria, the tiny cottage had a halo of starlight. Within, a God's eye and Huichol yarn paintings hung from a low roof, while an antique vacuum pump beat softly against the mountain silence. Old iron racks and lab benches held elaborate micro-glassware for reactions of hundreds of milligrams.

Sasha's lab was licensed by DEA and the State of California, so there was no fear of arrest. Doing thin-layer chromatograms to analyze new substances or monitor reactions-in-progress, Sasha's hands moved with refinement and accuracy as he described in muted tones the molecular mechanisms from which new medicines arose.

After Sasha first tested a new substance upon himself to assess its psychoactivity—whether others might benefit from the experience—a circle of trusted friends, many with advanced degrees, then would try it together in a special session. Rules were set: not to leave during the experience, not to forget to record one's recollections of the event. Through these courageous means, new probes to extend neuropharmacology were discovered.

My paltry skills were nothing beside his own. He sometimes spoke of the major acid chemists of yore, all of whom he seemed to know.

I once asked, "Why don't you oversee a team of foreign chemists, or use combinatorial techniques to generate thousands of novel structures, then assay them for psychoactivity with high-throughput screening?"

"Oh," he replied. "but thinking of new molecules—then creating them—are the greatest delights. One would miss all the *fun*."

Sasha was born to the manor of chemistry, not simply educated. There will be no other like him.

At these moonlit gatherings, I would see nearby Sasha's silver fingers raised high, and heard about him the silver waves of gentle banter. Those beguiled by his legend had found—to their delight—that he was playful and human. Bits of the sky stuck to him, and the birds seemed to sing through him. Those many who loved him often thought of Sasha, and of the next experimental substances from his everlasting river of thought and skill.

Suddenly, there he was. Aggregating and introducing people rather than isolating them, moving about with invitations.

"And your wife and little one? You must bring them again. So burdened with studies?"

To the delight of her many suitors, Wendy's hair was caught in a golden mesh, while Sasha and Ann in their rustic setting were hosts in some wild villa, and even the dear planets themselves those nights were aloft. The pulse of it all was not just from us, but from a latticework of information, bravery and vision flowing in from thousands of our extended family over the earth. Such was the complexity of things.

SASHA, beyond his intellectual gifts and his pure love of the art, had fortunate circumstances and inclinations. He had no hunger for material objects; his simple needs were met. Their home, hand-made by his father, was the same cottage in which Sasha was born. The elevated acreage on the mountainside allowed friends to roam; their narrow quarter-mile driveway during gatherings was lined with cars. Named "Shulgin Road" on maps, the county had acceded to his gleeful request.

There were no luxuries other than excellence of thought and accomplishment, loving minds and floating hearts and lingering twilights in the amorous summers, and the chirping of visiting children running all about, blithe as larks.

At Sasha and Ann's little dining table, just off the impossible kitchen, we would converse late into the evening. It was like some humble, privileged family—we felt among the chosen. I would look at the countenances of those assembled, for in these moments we at last belonged, respected and welcomed and understood as those who had seen too much. The faces, it seemed sometimes, were those of poor children around their first Christmas tree.

SO MANY IMAGES remain, memories of Sasha and Ann around the world. Buoyant exchanges in Amsterdam's cafés, the delight of neuroscientists at Harvard Medical School as Sasha lectured on entheogens with an aside about his freshman chemistry explosion in Harvard Yard. Sasha with Albert Hofmann and surrounded by scientists, students and youths with green Mohawks at 2 AM at European conferences on consciousness, picnics on Mt. Diablo with soft spring winds, Sasha looming with beatific mien after his presentations at American Chemical Society meetings, and always the magical nights, the unearthly flute music.

In Big Sur, at a small white frame house on a cliff over the Pacific, on a lawn with rare little Chinese apple and sour plum trees, the evening air the color of sunset, the mornings of bright showers shrouded in blue. Sasha discussing music as audible mathematics, cushions all about for a small circle of prominent figures: Czech psychiatrists, economists, a future Nobelist, professors of public policy and neuroscience from Harvard and Berkeley, a RAND thermonuclear war strategist, and Sasha always the prophesying oracle, settling under his hand our intellectual disarray and excitement.

In his talks and writings and among others, Sasha avoided any theological note in his references to the effects of newly created molecules. Full of energy, there was a touch of bustle about him. He had a horror of pomposity, wearing the same simple coat he'd had for years, so that one saw only his heart and mind, bright always as the sun.

I remember well the last night we spoke.

"I'm blind now," he confided, although Sasha's extended family throughout the world long had known. "Galileo was blind the last ten years of his life," I reminded him.

Beloved by all that knew him, Sasha's life was truly blessed. Our last words together were not on chemistry, but on music.

"You could still play your balalaika."

His father was Russian, the fine triangular instrument a family heirloom. Sasha was ambulatory somewhat, due to the loving care of Ann and his Tibetan women caregivers, who daily massaged his legs and guided him to sit in his revered laboratory with friends and colleagues who came to honor him.

"How many balalaika strings are there?" he asked. "Six," I replied, uncertain (There are three).

"That's right," he said, for he never discouraged anyone by asserting his knowledge. Then he said his final words to me. It was the last time I would ever hear his voice.

"There are heavenly harmonics."

— William Leonard Pickard
Author, *The Rose of Paracelsus*
January 2021

AFTERWORD

"Alright class, let's review." These oft-repeated words generated dread in the hearts of many university students as they approached midterm exams, which is where we close this volume. But Sasha Shulgin's Nature of Drugs was no ordinary course. He inaugurated his lectures with the assertion that this class would be a Socratic conversation, a give and take with his students, where listening and engagement were encouraged over note taking and a focus on fact accumulation. His style mixed a combination of rapid-fire, matter-of-fact exposition with seemingly meandering asides, much like the musical form of the rondo, with major themes and digressions. His was a supremely nimble mind, and these digressions were packed with his own experiences, observations, and philosophy.

When these lectures were given, this country was in a wave of anti-drug paranoia and fear. Richard Nixon had inaugurated the modern "War on Drugs" seventeen years earlier, with passage of the Controlled Substances Act (CSA), criminalizing possession of cannabis, some opiates, and many psychedelics, declaring they had no medical value and a high potential for abuse and addiction. With the admission in 2016 by Nixon's Assistant for Domestic Affairs, John Ehrlichman, that this legislation was not based on medical or scientific evidence, but was instituted as a blunt weapon against African American activism and the anti-war movement of the Left, we now know the Act was exclusively political, untethered to actual harm or medical realities. In the Reagan administration, fear of cocaine use was perceived to be a security problem of national import. The "Just Say No" campaign descended from the White House. Experimentation with drugs was seen as decadent, dangerous, and incompatible with productive social life. Basic research in this period was effectively eliminated, and publication of human effects of known or new psychoactive drugs screeched to a halt. Some four decades of potential scientific inquiry into the mechanisms of action, and potential for therapeutic use of banned drugs were lost. The CSA persists to this day, and has caused incarceration of perhaps as much as 20% of the United States prison population,[1] with all the attendant negative impacts on families and society in general.

Further, in 1987 access to objective information about drugs, particularly those banned by the CSA, was challenging for the lay public. The internet did not exist, let alone computerized searching of scientific literature. Those with interests in this area could use university libraries if they were accessible, but gleaning relevant information was, to say the least, arduous. Dr. Shulgin stepped into this breach, offering students without hard science backgrounds a much-needed survey of drug pharmacology, weaving in commentary on social mores, law, and the human right to independently control their bodies and psychological well-being. It is notable that Sasha chose to accompany this course with Andrew Weil's *From Chocolate to Morphine*, a non-judgmental survey of drug effects, and factors influencing choice. Another of Dr. Weil's works, *The Natural Mind*, addresses the underlying motivation for drug use, and proposes that alteration of the conscious state is an innate drive, akin to appetite for food or sex, that is not confined to humans, but exists across the animal kingdom. Well worth reading.

So far, the lectures have introduced Dr Shulgin's definitional foundations for the course. What are drugs? Initially, of course, we view drugs as chemicals that the body is unfamiliar with and alter its functioning. However, are there substances or forces from outside the body that could be considered to be drugs? He introduces us to the idea that even physical energy, such as radiation, could be lumped in with the definition, when it can be used to combat a disease, as in radiotherapy for cancer. In these discussions, he draws the awareness of his students out of their preconceived notions and prejudices about these things we call drugs.

---

1 Sawyer, W., & Wagner, P. (2020, March 24). Mass Incarceration: The Whole Pie 2020. Retrieved from https://www.prisonpolicy.org/reports/pie2020.html

He introduces us to the basic functions of the body, that allow for different routes of drug administration (oral, by injection, inhalation, rectal), and discusses the disposition pathways once drugs enter the body. Metabolism, sequestration in tissue sinks, penetration to active sites, and elimination are introduced. All of this lays the groundwork for examination of specific classes of drugs, which will be presented in subsequent volumes of this series.

Shulgin's message in this course draws on his conviction that the use of drugs is inevitable in modern life. Drugs used in medical practice have saved countless lives and eased suffering for millions. And, despite demonization by moralists, drugs that alter consciousness have had, and will continue to have attraction for explorers of the mind. His proposition is that the realization and acceptance of this attraction is vital, and the only sure way to extract benefit from the use of drugs is to support and precede their use with factual information about drug actions and potential benefits, as well as their risks.

Warts and All!

— Paul F. Daley, PhD
Alexander Shulgin Research Institute
February 2021

FOREWORD

Agar, M. (1985). Folks and professionals: Different models for the interpretation of drug use. *International Journal of the Addictions, 20*(1), 173-182.

Beck J. (1998) 100 years of "Just Say No" versus "Just Say Know." Reevaluating drug education goals for the coming century. *Evaluation Review 22*(1):15-45. doi: 10.1177 /0193841X9802200102. PMID: 10183299.

Bennett, W. (1989a). *National drug control strategy.* Washington, D.C.: Office of National Drug Control Policy.

Engs, R., & Fors, S. (1988). Drug abuse hysteria: The challenge of keeping perspective. *Journal of School Health, 58*(1), 26-28.

Forster, B., & Salloway, J. (1990). *The socio-cultural matrix of alcohol and drug use.* New York: Edwin Mellin Press.

Geber, B. (1969). Non-dependent drug use. In H. Steinberg (Ed.), *Scientific basis of drug dependence,* (pp. 375-393). London: Churchill.

Goode, E. (1993). *Drugs in American society.* (4th ed.). New York: McGraw Hill.

Pittman, D., & Staudenmeier, W. (1994). Twentieth century wars on alcohol and other drugs in the United States. In P. J. Venturelli (Ed.), *Drug use in America,* Boston: Jones and Bartlett.

Reuter, P. (1992). Hawks ascendant. Daedalus, 121(3), 15-52.

Reinarman, C. (1988). The social construction of an alcohol problem. *Theory and Society, 17,* 91-120.

Robins, L. N. (1993). Vietnam veterans' rapid recovery from heroin addiction: A fluke or normal expectation? *Addiction, 88,* 1041-1054.

Rosenbaum M. (1998) "Just Say Know" to teenagers and marijuana. *Journal of Psychoactive Drugs 30*(2):197-203. doi: 10.1080/02791072.1998.10399690. PMID: 9692382.

Skolnick, J. H. (1990). A critical look at the National Drug Control Strategy. *Yale Law and Policy Review, 8*(1), 75-116.

Waldorf, D., Reinarman, C., & Murphy, S. (1991). *Cocaine changes: The experience of using and quitting.* Philadelphia: Temple University Press.

Willis, H. (1989, September 9). "Hell No, I Won't Go." *Village Voice,* pp. 29.

Zinberg, N. (1984a). *Drug, set, and setting.* New Haven: Yale University.

Zinberg, N. (1984b). The social dilemma of the development of a policy in intoxicant use. In T. H. Murray, W. Gaylin, & R. Macklin (Eds.), *Feeling good and doing better,* (pp. 27-48). Clifton, NJ: Humana.

Zinberg, N., Harding, W., & Apsler, R. (1978). What is drug abuse? *Journal of Drug Issues, 8,* 9-35.

AFTERWORD

1 Sawyer, W., & Wagner, P. (2020, March 24). Mass Incarceration: The Whole Pie 2020. Retrieved from https://www.prisonpolicy.org/reports/pie2020.html

INDEX

5-HT 183, 189
5-Hydroxytryptamine 183, 189

A

Accidents 33, 263
Acetylcholine (ACh) 180, 183, 186–189, 230, 231
Acetylcholine Esterase (AChE) 187
Acetylcholine Esterase Inhibitors (AChEI) 187
Acute 42, 44, 63
Addiction 14, 40, 43, 63, 86, 212, 298–299, 329, 333
Addictive 13, 42, 63
Adrenaline 173–174, 180, 183
Adrenergic 180, 184, 186–189, 230
Afferent Signals 164, 167, 181
Agonist 43
Agriculture 23, 72, 74
Alchemy 76–77
Alcohol 4, 7, 32–33, 39, 44, 78, 121, 151, 158–160, 209, 209–211, 219, 228
Alkaloid 7–8, 80, 263
Amnesia 45, 207, 228, 231
Amobarbital 212
Amphetamine 155, 175–276, 223, 290
Analgesia 21
Anaphylaxis 42
Anesthetic 21, 95, 120–121, 161–162, 207, 209, 220
Aneurysm 139, 203–204
Animal Pharmacology 284
Antagonist 43
Arecoline 7–8, 10–11

Areflexive Mydriasis 176
Arterial Injection 115–118, 126, 176
Arteries 113, 116, 118, 123–126, 146, 169
Atrium 124–125
Atropine 144, 175–176, 230
Auditory Distortions 48
Autonomic Nervous System (ANS) 167–172, 176, 183, 187, 230

B

Babinski 201–202
Bactericide 206
Bacteriostat 206
Barbiturates 43–44, 88, 160, 211, 227, 298
Belladonna 171, 175–176, 248
Benzodiazepine 38, 211, 288
Betel Leaf (Betel Nut) 5, 8, 10
Biotransformation 126, 128, 136–137, 163
Blood 16, 28, 71–72, 112–119, 122–153, 166–169, 185–187, 220, 230, 263, 293 295–296
Blood Pressure 128, 167–170, 187, 203–205, 230, 290
Blood-Brain Barrier 164, 185, 231
Body Language 95
Bolus 114, 130, 136, 142, 146–147, 155–156, 295
Brain 12, 22, 28, 113, 114, 118–119, 124–128, 154–157, 161–170, 181–189, 227, 256, 296
Brainwash 253–255
Bulimia 226

Bureau of Drug Abuse Control 87–88
Bureau of Narcotics (BN)
Bureau of Narcotics and Dangerous
  Drugs (BNDD) 87

C

Caffeine 2, 7, 9, 145–147, 157, 202, 226
Cancer 8-9, 11, 26-29, 36, 88, 330
Capillary Bed 117, 123, 162
Carbamate Insecticides 188
Carcinogenic 26–27, 29, 36–37, 131
Cerebral Vascular Accident (CVA) 118,
  169, 203
Central Nervous System (CNS) 11, 164,
  172, 189, 227, 231
Cerebral Vascular Accident (CVA) 118,
  169, 203
Ceremonial Chemistry 38,
Chem Abstracts 271, 273, 280,
Chemotherapy 78–79, 81–82, 218
Chloramphenicol 293
Chloroform 221
Chlorpromazine 258–260, 299
Chocolate 2, 12, 20, 83, 330
Choice 16–18, 24, 33, 41, 148, 200, 236,
  245, 252–254, 330
Cholinergic 180, 187–189, 230
Cholinergic Synapse 187
Chronic 41–42, 63, 211, 229, 292
Cigarette 14, 83, 147, 152–153, 295, 297
Circulation 71, 117, 123, 143, 146,
  156–157, 200, 202, 227, 230, 297
Citation Index 278–280
Clearance 144–145
Clinical Trials 292
Clonic 44
Cocaine 14, 30, 86–88, 116, 177, 187,
  267–268, 271, 276–277, 329
Cod Liver Oil 20, 22–23
Codeine 21

Coffee 4, 13, 16, 39, 83, 147, 150, 157,
  259, 266, 285
Collectiveness 261
Catechol-O-Methyl Transferase
  (COMT) 186
Consciousness 165, 221, 231, 237, 240,
  244–245, 253, 262–263, 326, 331
Constitution 17–18
Controlled Substance Analogues
  Enforcement Act 92
Controlled Substances Act In 1970 91
Convulsant 43, 225
Convulsions 43–44
Coordinate Areas 242
Crack 30, 32
Craving 41
Criminal Forfeiture 90
Cross-Tolerance 43

D

DA183, 186
Darvon 120, 287
Datura 176
Dopamine Beta-Hydroxylase (DBH)
  186
Dichlorodiphenyltrichloroethane
  (DDT) 132–133
De Minimis 29
Delaney Amendment
Delusion 38, 46, 48, 231, 255–257
Department of Defense Authorization
  Act of 1982 90
Dependence 40–41, 43, 45, 63, 333
Deposition 117, 131
Depressants 87–88, 93, 187, 210–211,
  223
Derivatives 91, 189
Designer Drug Bill 92
Detoxification 127
Dexamyl 212

Diabetes 199–200

Diastolic 169–170

Diethyl Ether 78

Dilantin 139–140, 158, 211–212

Dimoxamine (Ariadne) 284, 289, 299

Dioxyphenylalanine 186

Diphenylhydantoin (Phenytoin) 139

Disclaimers 94

Dispositional Tolerance 43

Distribution Phase 136

DOB 138–139

Dolophine 63–64

Dopamine 180, 183, 186, 189, 231, 262

Dopaminergic 180, 189, 205

Double-Blind 20, 28

Dr. Hippocrates 34

Dreams 70, 220, 257

Drug Absorption 122

Drug Abuse 12, 15, 29, 33–35, 40, 65

Drug Clearance 144-145

Drug Education 3, 9, 12, 215

Drug Enforcement Administration (DEA) 35, 90

Drug-Seeking Behavior 41

Drunk 77, 160–161, 221

Duration of Deposition 131

Dyes 81

E

Ecgonine 30

Ecstasy (MDMA) 66, 92, 215,

ED-50 45,

Efferent Signals 165

Electroconvulsive Therapy 186

Embryology 110

Emergency Scheduling Act Of 1984 92

Enchanted Forest 250

Enteron 114, 121

Enzyme Induction 211

Epinephrine 180, 186–187, 262

Equilibrium 131, 133, 136–137, 143

Ether 78, 95, 220–221, 228

Etheromania 221

Euphoria 7–8, 44

Euphorica 228

Excitantia 223

Excretion 127, 129–130, 137, 144–145, 152–153, 163

Exsanguinate 113

F

Fahrenheit 451 18

Fainting 122

Fantasy 46, 48, 231

Fat Solubility 132

Food and Drug Administration (FDA) 24, 25, 29 34, 87, 212, 290

Fentanyl 216–217

Fever 19, 206

Field Sobriety Test 33

Fight-Or-Flight 173

Final 19, 39, 136, 235–238, 254, 327

Forfeiture 90

Freebase 30

Freedom of Choice 18, 33, 236, 252, 253–254

Fungistat 206

G

G% (Gram Percent) 151, 158

Gamma-Amino Butyric Acid (GABA) 189

Galileo 67

Gas Chromatograph-Mass Spectrometer 30

GC-MS 29

Ginseng 274, 275, 278

Glucuronic Acid 128–129

Green Medicine

Gunpowder 179, 271, 272

H

Habituation 229

Haldol 4, 177, 205

Half-Life 132, 134–139, 143 146, 152, 202

Hallucination 46, 231, 253

Hallucinogens 46, 87, 189, 223, 228–229

Hangover 44, 160

Harrison Act 85

Hashimoto's Disease 198,

Hawkins, Paula 18

Heart 112–114, 122–125, 128, 137, 146, 158, 161, 169–171, 202, 205

Heffter, Arthur 221

Hepatic Portal System 122

Heroin 21, 43, 63–64, 85–88, 138, 218–219, 228

Holding Law 65

Homunculus 182–183

Hyperexcitability 43–44

Hypnosis/Hypnotism 13, 250–251

Hypnotica 227

Hypodermic 83, 218

I

Illness 15, 19, 22, 43, 69, 79–80, 169, 206, 226, 246–248, 252, 255–257

Illusions 46, 231

Imagery 46, 48, 169, 231, 253

Imipramine 187, 289

Immortality 77

Immunological Response 198

Implication 34, 275

In Entera 114

Increasing Ionizability 129

Inductive Inference 275–276

Inebriantia 223

Inference 275–277, 289

Inhalation 117, 126, 147, 331

Injection 114–118, 124, 126, 130, 136, 139, 144, 147–149, 295, 298

Insufflation 116

Insulin 28, 41, 167, 199–201, 209

Intensive Care Unit (ICU) 209

Interoceptives 167

Intestines 111, 115, 117, 121, 123–124, 126, 200

Intramuscular 116

Intraspinal 118

Intravenous 115–116, 124, 136, 144, 162, 220, 295

Investigational New Drug (IND) 290

Involuntary 41, 167–168, 177

Iodine 113, 198

Iodine-131 198

Ionized 121, 144

Ipecac 226–227, 273

Irrational Mixtures 212

J

Jimson Weed 176

K

Ketamine 165

Kinetics 141, 152, 158, 295

L

Law of Concentration 136

LD-50 45,

Lethal 36, 45, 150, 162, 211, 277

Lungs 117, 129, 130–131, 146–147, 202, 224

Lye 225

Lyse 230

M

Malathion 188

Mandragora 248

Mandrake 78, 248

Marijuana 4, 31–32, 78, 86–88, 132,

216, 218–219, 248
Mass Medication 24
Materia Medica 15, 81, 274
MDMA 92, 215
Melatonin 263
Memory 45, 72, 228, 242–245,
Mental Illness 48, 70, 240, 257,
    260–264
Meperidine (Demerol) 217, 228
Mercaptan 10–11
Merck Index 272–274, 280
Mesentery Net 117
Mesmer, Anton 246
Mesmerization 246
Meta-T-Butylphenol 282
Metabolic Tolerance 43
Metabolism 127, 140, 145, 147, 153–156,
    158, 161, 214, 288, 290
Metabolize 43, 128, 144, 157, 202, 263
Methadone 13, 64
Methamphetamine 155
Methylcholanthrene 27
Methylene Chloride 130–131
Midterm Discussion 19, 235–237, 329
Mixed Drugs 162, 211, 212
Mnemonic 44–45
Monoamine Oxidase (MAO) 186, 230
Morphine 2, 12, 18, 20–22, 43, 64,
    80, 83–86, 92, 95, 138, 216–217,
    228, 330
Muscle 44, 114–117, 122, 161, 166–168,
    171–173, 174–176, 183, 205, 227
Mutagenic 36–37
Mutations 36
Mydriasis 171, 175–176

N
Narcotic 35, 44, 88, 217
National Organization for the Reform
    of Marijuana Laws (NORML) 218

NE 183, 186
Needle Orientation 155
Nerve 160, 162–163, 165–166, 168,
    177–184, 186–187, 195–196, 230–231
Nerve Conduction 178
Neurotransmitters 161, 163, 179,
    181–184, 186, 195, 229–231, 262
Newborns 201
Nicotine 7–8, 144, 152–154, 156–157,
    295–297
Nitrous Oxide 95, 221
Nontoxic 35–36
Noradrenaline 180, 183
Norepinephrine 161–162, 180, 183–187,
    189, 230, 262
Note Taking 233–235, 329

O
Ontogeny 110
Opiate Antagonist 43
Opium 63–64, 78, 80, 83–87, 247–248
Opium Wars 84
Optical Isomers 91
Organic Chemistry 79, 81, 271, 319, 323
Oxidation 127

P
Pain 20–22, 25, 83–84, 118, 162, 181,
    206, 216, 228–229, 242–243,
    246–248, 250–252
Pancreas 167, 199–200, 205
Paranoid Delusion 255–257
Parasympathetic 170–172, 174–177,
    187–188, 282
Parasympatholytic 38, 171–172,
    174–176, 207, 230
Parasympathomimetic 172, 174,
    176, 230
Parathion 188
Parenteral 116–117, 121, 126, 131

Partition 151–152
Patents 11, 286–288, 322
Pavlovian Response 253
PCP 32, 128, 165, 219
PDR 294
Penicillin 82
Peripheral Nervous System (PNS) 164, 167, 186, 230
Phantastica 228, 231,
Pharmacodynamic Tolerance 43
Pharmacodynamics 109, 152, 154
Pharmacokinetics 109, 137, 157
Phenobarb 139–140, 160
Phenobarbital 211–212
Phenytoin 140
phocomelus
Phylogeny 110
Physical Dependence 40–43
Physician's Rights 85
Physostigmine 282
Pills 22, 89, 130
Pilocarpine 189
Pineal Gland 263,
Pink Spot of Schizophrenia 258
Placebo 20, 22, 74, 120
Plasma Half-life 135, 144
Plumbing of the Body 163
Poison 46, 75, 280
Poisonous 36, 201, 288
Polypharmacy 208–209
Portal 122–124, 156, 200, 296
Portal Systems 122
Postsynaptic 180, 187
Potentiation 210
Power 65–66, 236, 252–253
Pregnancy 37–38, 213, 215
Prematures 202
Presynaptic 180, 184, 186–187
Presynaptic Nerve Ending 180, 184, 186

Prodrug 137, 139
Prophylactic 11, 213
Prophylaxis 42
Proprioception 166–167
Proprioceptors 166
Psychedelics 12, 229, 329
Psychodysleptic(s) 229
Psychological Dependence 13, 41, 229
Public Law 85–929
Pulmonary 124, 126 139, 161
Pupil Size 171–176, 217
Pure Food and Drug Act 86
Pylorus 111, 121

R

Radial Muscles 171–175
Radioactive Marker 136
Radioactivity 23, 119, 134–136, 139, 141, 196, 198, 214
Rationalization 13, 35, 41, 43, 64
Receptor 128, 133, 162, 170, 179–181, 184, 186–187, 189
Receptor Site(s) 1, 22, 43, 112, 127–128, 149, 152, 159, 166
Recreational Use of Drugs 155
Redistribution 137, 144–145
Reflexive Mydriasis 175–176
Relapse 41
REM (rapid eye movement) 44–45
Renal Clearance 143
Renal Return 117
Research 219, 221, 255, 264–266, 268–279, 281–285, 293–95, 299, 320–322
Reserpine 169, 187
Retina 46, 130, 166

S

Saccharine 27–29
Salvarsan 81

Sarin 188
Scheduling 89, 91–92, 216,
Schistosomiasis 129
Schizophrenia 244, 252, 257–259
Schoenfeld, Gene 34
Scopolamine 171, 175–176, 228, 230
Secobarbital 159, 212
Sedative 38, 159, 212
Sedative-Hypnotic 37, 45, 227
Serotonergic Synapse 180
Serotonin 180, 183, 189, 231, 262–263
Sevin 188
Sex 3, 25, 38, 41, 149, 215, 274, 330
Shaman 69–72, 74, 79, 252, 320
Shock 117
Short-term Memory 228, 244
Sickness 69, 196, 258,
Skin Popping 116
Sleep 16, 25, 44, 48, 70, 73, 78, 159, 189,
    212, 227–229, 242, 250–251
Smoking 13–14, 33, 83, 85, 117, 147,
    152, 156, 218–220, 295–298
Snorting 116
Snuffs 155
Sodium Ascorbate (Vitamin C) 144
Sodium Hydroxide (Lye) 224–225
Somatic Homunculus 182
Sphincter Muscles 172, 174–176
Stars 72, 74–75, 77, 79, 93, 229, 247, 281
STAT 210
Stomach 20, 111, 115–116, 121,
    124, 126, 156–157, 160–161, 200,
    224–225
Strychnine 225–226
Stupefactant 78
Subcutaneous 116
Sublingual 114
Sugar 28, 122–123, 128–129, 156,
    167, 199–201, 286, 298
Sulfanilamides 81–82

Sulfhydryl 8
Sulfur 129–130
Sulfuric Acid 78, 221
Suppository 22, 112, 117
Sweat 129
Sympathetic 170–172, 174–177,
    186–188, 230, 282
Sympatholytic 172, 176–177, 205,
    230, 231
Sympathomimetic 172–177, 205, 230
Synapse 178–181, 183, 184, 186–187,
    189
Synaptic Cleft 179–181
Synergistic 210–211, 322
Systolic 169
Szasz, Dr. Thomas 38

T
Tabun 188
Tachyphylaxis 42
Talc 130
Tax Equity and Fiscal Responsibility
    Law of 1982 90
Tea 9, 14, 147, 176
Temporal Lobes 242
Teratogenic 46, 36, 38, 88
Terminal Half-life 136
Thalidomide 37
THC (Tetrahydrocannabinol) 132–133,
    217–219
Therapeutic Index (TI) 162
Thyroid 44, 112–113, 196, 198, 205
Tissue Depot 133, 144
Tissue Distribution 143, 147
Tobacco 4–5, 7–9, 11, 14, 33, 41, 78,
    82–83, 152, 156, 219, 295, 298
Tolerance 41–43, 63, 159
Tongue 114–116, 182
Tonic 44
Torus 110–111

Toxaphene 289
Toxic 35–36, 45–46, 82, 119, 127,
    149, 203
Toxicity 35, 81, 131, 162, 208, 289–290
Tranquilizers 4, 89, 177, 189, 258, 299
Transformation 14, 67, 69, 71, 126,
    139 263
Transmutation 76
Trematodes (blook flukes) 127
Tuinal 212, 273
Tyrosine 185

U

Ups, Downs, and Stars 93, 223, 228
Urine 31, 37, 129, 139, 144, 150,
    153–154, 170, 173, 208, 259–260
Urine Testing 31, 198

V

Veins 113, 115–116, 122–124, 126, 154,
    169, 203, 216
Venous System 113, 117
Ventricle 119, 124–125
Vesicles 186
Visceral Senses 166

Vitriolic Acid 78
Volume of Distribution 131–133, 152
Voluntary 167–168, 177
Vomiting 36, 43, 173, 207, 207, 223,
    224, 226

W

Weight Percent 150–151
Withdrawal 13, 41–45
World War II (WWII) 21, 63, 81, 82,
    87, 233

X

X-rays 20, 119, 132, 138
X-Y Axis 134,
Xylocaine 161–162

Y

Yage 155

Z

Zectran 188 283